# BIOMASS AND ALTERNATE FUEL SYSTEMS

# BIOMASS AND ALTERNATE FUEL SYSTEMS

## An Engineering and Economic Guide

Edited by

**THOMAS F. McGOWAN**

Associate Editors

**MICHAEL L. BROWN**
**WILLIAM S. BULPITT**
**JAMES L. WALSH, JR.**

AIChE®

A JOHN WILEY & SONS, INC., PUBLICATION

A Joint Publication of the American Institute of Chemical Engineers, Inc. and John Wiley & Sons, Inc.

Published by John Wiley & Sons, Inc., Hoboken, New Jersey
Published simultaneously in Canada.

For general information on our other products and services please contact our Customer Care Department within the United States at (800) 762-2974, outside the United States at (317) 572-3993 or fax (317) 572-4002.

Wiley also publishes its books in a variety of electronic formats. Some content that appears in print, however, may not be available in electronic formats. For more information about Wiley and AIChE products, visit our web site at www.wiley.com.

*Library of Congress Cataloging-in-Publication Data is available.*

McGowan, Tom, 1950–
Biomass and alternate fuel systems : an engineering and economic guide / Thomas F. McGowan.
    p. cm.
    Includes index.
    ISBN 978-0-470-41028-8 (cloth)
    1. Biomass energy—Economic aspects. 2. Biomass energy—Environmental aspects. 3. Renewable energy sources—Economic aspects. 4. Renewable energy sources—Environmental aspects. I. Title.
    TP339.M38 2009
    662'.88—dc22                                                                    2008035476

Printed in the United States of America.

10 9 8 7 6 5 4

# CONTENTS

*Preface*                                                                    vii
*Acknowledgments*                                                             ix

**CHAPTER 1** Introduction to Alternate Fuels                                  1

**CHAPTER 2** Fuel Properties and Combustion Theory                           13

**CHAPTER 3** Liquid Fuels from Biomass                                       37

**CHAPTER 4** Biomass Combustion Equipment—Steam, Hot Oil,                    45
and Hot Gas

**CHAPTER 5** Biomass Fuel Storage and Handling                               95

**CHAPTER 6** Cogeneration and Power Generation                              125

**CHAPTER 7** Emissions and Control                                          137

**CHAPTER 8** Environment and Safety: Rules, Regulations, and                157
Safe Practice

**CHAPTER 9** Biomass Fuel Supply and Purchasing                             171

**CHAPTER 10** Fuel-Switching Feasibility Study Methodology                  175

**CHAPTER 11** Economic Analysis of Biomass Combustion Systems               183

**CHAPTER 12** Biomass Fuel Processing Routes and Economics                  191

**CHAPTER 13** Biomass Fuel Processing Network                               221

**CHAPTER 14** Example Feasibility Study: Nonforest                          225
Products Facility

**APPENDIX 1** Equipment Manufacturers/Vendors Listing                       239

**APPENDIX 2** State Forestry Commission Offices                             247

**APPENDIX 3** Glossary                                                      253

**INDEX**                                                                    259

# PREFACE

$W$ E ARE ONCE AGAIN IN an energy and fuels crisis, with costs escalating dramatically and supplies tight. This has happened before and, once again, the economy and industry will adjust to new conditions.

This book is about how to adjust to this difficult situation by using renewables for industrial applications, while cutting operating costs. In addition to providing economic benefits, the switch to renewable wood and agricultural fuels reduces greenhouse gas emissions, as these renewables are assigned a zero greenhouse emission factor and are considered "carbon neutral."

Some of the material in this book is from *The Industrial Wood Energy Handbook,* written in 1984. This book is an update and expansion of that work, adding information and broadening the scope to include agricultural feedstocks and the use and production of liquid fuels such as ethanol from cellulosic (noncorn) feedstocks. It also covers fireside slag treatment (not a small issue in wood and agricultural combustion systems!), has up-to-date equipment vendor information, more detail on the furnace grate systems that are the heart of many biomass and wood combustors, and information on hot oil systems and hot air, introductory material about sustainable biomass yields, as well as the basics of greenhouse gas emissions calculation methods.

You will find this book to be useful in choosing the right equipment the first time, producing feasibility studies that define project economics, and cutting industrial fuel costs while reducing greenhouse gas emissions.

THOMAS F. McGOWAN

*Atlanta, Georgia*
*January 2009*

# ACKNOWLEDGMENTS

Major contributors to this handbook are:

Thomas F. McGowan, PE, Editor, Primary author, Chapter 1 on alternate fuels, Chapter 4 on slag

William S. Bulpitt, P.E., Associate Editor, Chapter 2 on biomass sustainability

Michael L. Brown, PE, Associate Editor, Chapter 12 on wood pellet production

James L. Walsh, Jr., Associate Editor, Chapter 4 on equipment vendors

Dennis Coughlin, Doug McLain, and Craig Smith, Chapter 1 contributors on alternate fuels

Tom Wechsler, Chapter 1 contributor on alternate fuels, Chapter 4 on hot oil and hot air systems and grates

Jonathan Scurlock, Chapter 2 on fuel properties and combustion theory

Ryan Melsert, Chapter 3 on liquid fuels, Chapter 11 on economic analysis, appendices 1 and 2 updates, book production

We are grateful for the assistance of the following in preparation of this handbook: Haeja Han, Publisher and Executive Editor, Technical Publications, AIChE; Robert Esposito, Wiley Publishing; John Wells and Devon Dartnell, Georgia Forestry Commission; and Sam McGowan, graphics.

## Acknowledgements from *The Industrial Wood Energy Handbook*

Major Contributors to the 1984 *Industrial Wood Energy Handbook*:
William S. Bulpitt, Grant B. Curtis, Jr., Steven J. Drucker, Michael L. Brown, Robert J. Didocha, Thomas F. McGowan, James L. Walsh, Robert D. Atkins, and Dr. Badarinath S. Dixit

The following assisted in preparation of the 1984 *Industrial Wood Energy Handbook:*

John C. Adams, Jr., Michael S. Smith, David E. Harris, F. Dee Bryson, Anthony D. Jape, Douglas Davis, Bryan Miller, William Hartrampf III, Joseph Saucier, David Lapin, Dr. Arthur Shavit, David Pugmire, and Joanne Bocek

The 1984 *Industrial Wood Energy Handbook* was written under Contract Number DE-FG05-79-ET 23076, which was funded by the U.S. Department of Energy.

# CHAPTER 1

# INTRODUCTION TO ALTERNATE FUELS

## INTRODUCTION

Fuel costs rose sharply in 2005 due to hurricane damage to off-shore platforms and refineries in the Gulf of Mexico. This was followed by a drop in the strength of the U.S. in dollar in 2007 and 2008, international developments in the oil supply, and increased demand. Competing fuels—coal and natural gas—rose in price in lockstep. Natural gas has risen in cost, and has been reported by clients to be as much as $17/MM Btu for industrial use in the northeast, and the cost of a 42 gallon barrel of oil hit $140, which is $3.33 per gallon just for the raw material.

Operating firms with fuel bills that are a high percentage of their costs are seeing profit margins erode, and one solution to the problem is fuel switching and use of alternate, sometimes over-looked, lower cost fuels.

There are many alternate fuels available to replace the big three—gas, oil, and coal. These include biomass, in particular, wood, but also including bagasse, fast growing switch grass, agricultural feedstocks, and used cooking oil. Other alternate fuels are covered in this book as a point of comparison and in less detail than biomass, and also because they may be cofired with biomass fuels. They include used/recycled oil, tires, and solid waste.

Parts of this chapter were excerpted by special permission from *Chemical Engineering,* March 2006. Copyright © 2006 by Access Intelligence, New York, NY 10038.

Some combustion systems were set up for multifuel firing and can be easily changed over, whereas others can be refitted and adapted to new fuels.

Heat recovery is the other option—adding heat exchangers to current processes, or changing over to new types of equipment, for example, from an elevated flare to a thermal oxidizer with a heat recovery boiler to recoup useful heat. This approach can be used to cut greenhouse gas emissions while saving on fuel costs.

This section covers the heating value, costs, and other properties of alternate fuels and fossil fuels, the changes required to fire alternate fuels, and options for heat recovery. It also addresses in brief the regulatory aspects of fuel switching and impacts on emissions. In addition to information on the *use* of alternate fuels, this section also covers *production* of secondary fuels from biomass feedstocks, for example, ethanol from cellulosic feedstocks and transportation fuels from agricultural feedstocks.

## WHAT FUELS ARE USED?

Conventional fossil fuels rule the marketplace. They include natural gas, propane, fuel oil, and coal. Alternate fuels include:

- Wood and other biomass and agricultural feedstocks
- Reclaimed oil
- Petroleum coke
- Solid wastes
- Biogas (methane from bioprocesses, used on-site)
- Used tires
- Used cooking oil
- Ethanol from wood or agricultural feedstocks
- Plant and refinery gas (used on-site)

Table 1-1 shows heating values and costs for industrial use of alternate fuels and selected fossil fuels, which are the benchmark for comparison on prices and heating values. Establishing a fuel price is not as simple as it might seem, as database figures must be adjusted to match the application, location, and time of year, and taxes, transport, broker fees, and market must be taken into account to come up with representative prices paid by the end user. The type of contract, for example, interruptible service for natural

**Table 1-1.** Heating values and costs for fuels

| Fuel | Gross heating value (Btu/lb) | Approximate cost ($/MM Btu) | Comment (source, date) |
|---|---|---|---|
| Natural gas | 23,896 | $7.76 | Industrial supply (EIA 2007) |
| Propane | 21,523 | $12.51 | $1.13/gallon, wholesale (EIA, 2007) |
| No. 2 fuel oil | 19,567 | $14.30 | $2.01/gallon, wholesale (EIA, 2007) |
| No. 6 fuel oil | 18,266 | $7.89 | $1.25/gal, sale to end user; price varies with sulfur content (EIA, 10/07) |
| Coal | 9,000–15,000 | $1.78 | $39/ton; cost based on 11,000 Btu/lb. U.S. industrial delivered average (EIA 2007) |
| Wood waste | ~4,250 at 50% mc | $0.90 | $9–10/ton, delivered (Ref. Timber Mart South, third qtr 2005) |
| Wood, whole tree chips | ~4,250 at 50% mc | $3.20 | Mid to high $20s per ton (Ref. Timber Mart South, third qtr 2005) |
| Wood, tub grind waste material | ~4,250 at 50% mc | $2.60 | Low to mid $20s per ton (Ref. Timber Mart South, third qtr 2005) |
| Wood planer shavings | ~7,650 at 10% mc | NA | Dry planer shavings, sander dust. Light and hard to transport; mainly used on-site |
| Used/recycled oil | ~17,500 | $9.50 | $1.20/gal, or 70% of No. 2 oil based on cost/gallon. Derived from engine lube, cutting oil, etc. Typically, viscosity is No. 4 fuel oil equivalent. Density ~7.2 lb/gal |
| Pet coke | 15,250 dry; 14,200 wet | $1.29–$1.94 | $36/ton (EIA 2007); $50/ton (FERC 2007) Pet coke; varies in volatile content and heating value. Sulfur 4–6.5% (EIA), Ash ~ 1% |
| Tires | 15,500 (metal free) | $1.61–$3.23 | $50–100/ton 96% metal free 2″ top size. TDF chips (RB, 1/11/08). Sulfur content ~1.2%; contains ash from metal and glass/polyaramid cords |

*(continued)*

**Table 1-1.** Heating values and costs for fuels

| Fuel | Gross heating value (Btu/lb) | Approximate cost ($/MM Btu) | Comment (source, date) |
|---|---|---|---|
| Used cooking oil | 16,900–18,500 | Local price variation, at or below #2 fuel oil | Need to decant water, filter grit. Price varies with time of year and local supply/demand |
| Ethanol | 13,161 | NA See at right | Ethanol from cellulosic feed stocks is in development via several routes; market prices do not exist |
| Biogas | Varies | NA | Biogas is typically 500–900 Btu/ft3, depending on source |
| Plant gas | Varies | NA | Heating value varies with upstream process and diluents, if any |
| Type 0 Trash | 8500 | NA | Incinerator Waste Standard |
| Type 1 Rubbish | 6500 | NA | Incinerator Waste Standard |
| Type 2 Refuse | 4300 | NA | Incinerator Waste Standard |
| Type 3 Garbage | 2500 | NA | Incinerator Waste Standard |
| Type 4 Human and animal remains | 1000 | NA | Incinerator Waste Standard |
| Type 5 By-product waste, liquid, solid, sludge | By test | NA | Incinerator Waste Standard |
| Type 6 Solid by product industrial waste | By test | NA | Incinerator Waste Standard |

Notes: EIA, http://www.eia.doe.gov/; Timber Mart South/UGA, Tom Harris, 12/6/05; RB Rubber Products, Pete Daly, 503-283-2261, Portland, OR.

gas, also affects final price. Many databases show current, 30 day spot pricing. An example is the NYMEX (NY Mercantile Exchange), which showed natural gas in the $15/MM Btu range on 12/8/05. This price is based on the Henry Hub in Louisiana, and does not include delivery and markup. This is a higher price than that found in longer-term contracts, which tend to smooth out short-term spikes and dips. Based on EIA data, the Henry Hub price for natural gas went up 63% from January 04 to January of

2009 (projected price). For more on forecasting and market movements see www.eia.doe.gov/emeu/steo/pub/contents.html, and for relationships between gas and oil see www.dallasfed.org/research/swe/2005/swe0504c.html.

Biogas (from wastewater treatment, conversion of manure, and from biological degradation in landfills) and plant and refinery gas are confined to use on-site or a short distance off-site and are not transportable. Setting a price for these fuels is difficult, as some of the cost may be attributable to regulatory requirements (e.g., regulations on landfill gas), whereas the rest might be allocated as internal cost to the end use. For example, most landfill gas must be collected and flared if it will not be used as a fuel for an engine or furnace. So the cost allocated for use as a fuel might be only that of any extra gas cleanup and extra fan or blower power and piping required to transport the biogas the extra distance to the engine.

Reclaimers throughout the United States produce used oil. It is frequently referred to as "spec oil," referencing U.S. Environmental Protection Agency (EPA) terminology that allows its use as a product and takes it out of the waste category. See CRF 279.10 for more detail (http://ecfr.gpoaccess.gov, part 40, subparts 266–299) and Table 1-3 for EPA requirements. This EPA regulation also allows reuse of used oil on-site by a generator (e.g., blending with diesel fuel for use in the company's vehicles), and under these conditions, the fuel is not subject to EPA used oil or waste regulations. Typical properties for used "spec oil" can be found in Table 1-2. The asphalt industry and industrial and utility boilers are major users of used "spec oil."

With the exception of some cement kilns that feed whole tires, TDF (tire-derived fuel) is normally burned as chips in boilers and cement kilns. Prices range widely, based on state used-tire tar-

**Table 1-2.** Typical properties of used oil from reclaimer

| Parameter | Value |
| --- | --- |
| Heating value | >135,000 Btu/gal |
| Water | <1% |
| Solids | <1% |
| Viscosity | 90–120 SSU at 100°F |
| Flash point | >100°F |

Source: Perma-Fix, Inc., personal communication, Doug McLain, 12/5/05.

**Table 1-3.** EPA 40 CFR 279.11, limits for used oil specification level[1,2]

| Parameter | Value |
|---|---|
| Arsenic | ≤5 ppm |
| Cadmium | 2 ppm |
| Chromium | 10 ppm |
| Lead | 100 ppm |
| Flash point | ≥100 °F |
| Total halogens | 4000 ppm[3] |

1. Used oil not exceeding any specification level is not subject to this part (40 CFR 279) when burned for energy recovery. The specification does not apply to mixtures of used oil and hazardous waste that continue to be regulated as hazardous waste [see § 279.10(b)].
2. Applicable standards for the burning of used oil containing PCBs are imposed by 40 CFR 761.20(e).
3. Used oil containing more than 1,000 ppm total halogens is presumed to be a hazardous waste under the rebuttable presumption provided under § 279.10(b)(1). Such used oil is subject to subpart H of part 266 of this chapter rather than this part when burned for energy recovery, unless the presumption of mixing can be successfully rebutted.

iffs, number of processors, and supply/demand. TDF prices are generally similar to coal.

Coal and coke can be used in industrial equipment, and both have been fired in cement kilns [1]. Pet coke (petroleum coke) prices are in the range of that of coal. Shipping dictates that there is a price advantage for users near the refineries and coke plants that produce this material as a byproduct of heavy-oil refining. It should be noted that this material tends to be high in sulfur content, and any combustion system using it needs to have equipment capable of dealing with the $SO_2$ emissions. More detail on pet coke can be found in the referenced EPA report [2].

Biomass fuels are becoming a more common alternative fuel source to fossil fuels as conventional energy prices rise. The primary biomass fuel is wood waste; however, bagasse (sugar cane residue), ag-fuel, or fuels derived from straw, rice hulls, and shell hulls, biomass grown as a fuel crop (e.g., switch grass) and other agricultural sources are becoming more popular. Plywood, lumber, OSB (oriented strand board), and related plants have long used bark, wood waste, planer shavings, and sander dust from plant operations to provide process heat for drying and pressing of boards rather than fossil fuels. However, biomass fuels are finding applications in nonforest products industries as well. Biomass fuels are used to replace up to 5–10% of coal during cofiring in some utility boilers at power plants. The reason for replacing only a portion of the fossil fuel is that generally wood fuels have a lower heating value and higher moisture content, so that only a portion of coal fuel can be replaced without significant loss of boiler performance and output.

Advantages to biomass are lower fuel cost and lower emissions, since wood fuels are much lower in sulfur content than most typical coals. They also contain ash, which has alkali components that can react with and remove some sulfur dioxide. In addition, in those countries that have ratified the Kyoto Protocol, the use of naturally derived fuels as opposed to fossil fuels is a common method to reduce the emission of "greenhouse gases" and obtain CO2 emission credits, since renewable fuels from plant sources are considered "CO2 neutral" under this accord. This is common in Europe and is an emerging market in itself, with wood pellet producers in the United States shipping their product across the Atlantic. Fossil fuels are burdened with associated $CO_2$ greenhouse gas emissions. Although company-wide studies can cost thousands of dollars to count direct and indirect $CO_2$ emissions, when it comes to burning a fuel, the math is really quite simple: Multiply the fuel weight per year by the carbon fraction of the fuel, then multiply by 3.66 (3.66 is the molecular weight of $CO_2$ divided by that of carbon) and divide by 2000. This result is the tpy (tons/yr) of $CO_2$ generated. The carbon fraction in the fuel can be found in reference texts or via lab work. Basic chemistry can also be used to calculate the carbon fraction. For example, for methane, the carbon weight is 12, when divided by the total weight of 16, the result is 0.75, or 75% carbon.

## COMBUSTION ISSUES AND APPLICATION TO EQUIPMENT

Burning alternate fuels requires going back to the basics. We have to know fuel properties, as the chemistry of the fuel dictates the final stack-gas flow and products of combustion. For example, higher sulfur will mean more $SO_2$ generated, and for some fuels, such as wet wood hog fuel, higher excess air (in the range of 50%) is required using air swept stokers, much more than for oil and gas, which are usually in the 15–25% range, or lower when $O_2$ trim or mass flow control is used.

## Application to Equipment

Some equipment is easy to alter for firing of alternate fuels. For example, boilers set up for natural gas are also sold with gas/oil firing systems. Installation of oil tankage, fuel trains and atomizers, and substitution of an oil gun for the gas spud may be all it takes to change over. Changing from No. 2 fuel oil to heavy oil or reclaimed used oil will require the addition of steam or compressed-air atomizers and oil-heating equipment.

Changing to solid fuel is much more difficult. For example, compact boilers set up for gas/oil are simply not compatible with solid fuel firing, unless a gasifier is used to convert the fuel upstream. Very large dryers and calciners are more amenable to such a change, as they were designed to handle particulate matter, and have pollution control systems that remove particulates. Some manufacturers are working on a new generation of mid-sized (100–200 MM Btu/hr) pulverized coal burners to respond to emerging markets and applications that can tolerate particulates.

For larger, long-term commitment to solid fuels, the best fix may be the purchase of new equipment designed for its use.

Beware that some elements in fuels can cause major problems even in small amounts. These include sodium and potassium that can form low melting point ash, and sulfur and chlorine that form acid gases. An additional issue is fuel bound nitrogen, which forms $NO_x$. These problems can cause equipment outages due to fouling and corrosion, and raise emissions.

One issue with biomass fuels is their alkali ash compounds, which form salts in the combustion process and/or may lower the ash softening points, causing fireside fouling problems. This is particularly true with agricultural biomass, while wood-based biomass has lower alkali concentrations when compared to many agricultural feedstocks. Still, for any biomass fuel, it is advisable to analyze for alkalis, and make a determination of the fuel's fouling and slagging potential. Also, any heat recovery equipment must be properly designed to minimize the potential any of these fuels have for fouling. Fireside additives have been used, generally with good effect, to reduce fouling problems and slagging of grates.

## $SO_2$, Acid Gas Dew Point, and Heat Recuperation

Heat recuperation is used for many processes, and more of these systems are being added today to reduce fuel costs. Higher sulfur fuels can harm heat recuperators, as heat recuperation lowers stack-gas temperatures and can cause dewpoint condensation of acid on heat exchanger surfaces, causing rapid corrosion. This is particularly an issue for air preheaters and economizers used to preheat boiler feed water via heat exchange with stack gases. Figure 1-1 addresses this subject and will help keep you out of trouble. It provides the acid-gas dewpoint based on sulfur loading in the fuel.

For example, a fuel with 20 grains (there are 7000 grains per lb) of sulfur per 100 ft³ of fuel (fuel gas) divided by the fuel heating value in Btu/ft³ of the fuel gas will have an acid-gas dewpoint of 242°F at 20% excess air. Heat exchanger wall temperatures below that will result in liquid acid on the metal surface, and heat exchangers will experience high rates of corrosion. Graphs of acid-gas dewpoint are available in many references, such as ASHRAE's *Handbook of Fundamentals* [4].

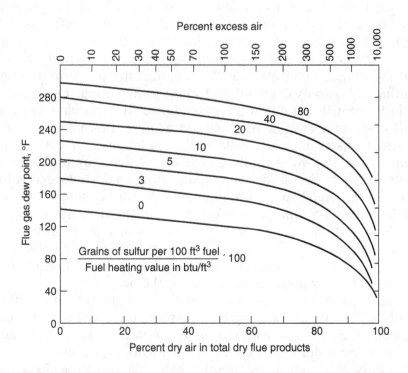

**Figure 1-1.** Influence of sulfur oxides on flue-gas dewpoint. (Chart © ASHRAE, *Handbook of Fundamentals,* 1993, used with permission ASHRAE, www. ashrae.org.)

## ABOUT EMISSIONS AND REGULATIONS

In brief, the regulatory problems involved in fuel switching are:

- Sulfur content/$SO_2$ emissions
- Ash content/particulate emissions
- Chlorine content/HCl and $Cl_2$ emissions
- Nitrogen content/$NO_x$ emissions

If the original permit allowed firing of other fuels (e.g., No. 6 oil as alternate to natural gas), then no repermitting is required to switch to fuel oil. However, it would be worth checking the permit application, the operating permit, any language in the facility's State Implementation Plan (SIP) permit or Title V permit (number of hours of operation may have been limited to reduce $NO_x$) that may need to be modified, thus requiring the facility's operating permit to be reopened and subsequently reissued.

Furthermore, depending on whether the area in which the facility is located is in attainment or nonattainment with the national ambient air quality standards (NAAQS), coupled with the magnitude of the emissions increase that may occur as a consequence of the fuel switching, New Source Review (NSR) permitting applicability could be triggered. This could result in additional permitting complexity (beyond just simply reopening the facility's Title V permit) that could involve additional requirements to install postcombustion controls and/or acquire internal or external emission offsets in order to permit the modification. Thus, the important first step in evaluating any project that may involve fuel switching is to first undertake a permitting analysis to assess the underlying regulatory requirements the facility is subject to, and establish the optimum path-forward permitting strategy.

Some alternate fuels emit less pollution. For example, tests with used cooking oil firing boilers showed lower CO and $NO_x$ emissions than with fuel oil. Thus, substituting this alternate fuel may require only notification to the regulators.

In the end, your relationship with the regulators is a critical element. If you have an ongoing, positive relationship, you may find that it may not take all that much paperwork to change fuels, as long as emissions do not rise past regulatory thresholds.

EPA's AP-42 guide to emission rates may prove useful when exploring fuel switching. It contains a compilation of air pollution estimates for industrial stationary sources and other air pollution sources. It is available online, and provides emissions for various applications (e.g., boilers and cement kilns), by industry and by fuel. Useful URLs can be found in reference [4].

Note that the above discussion applies to fuels and nonhazardous waste. It does not apply to firing of RCRA "hazardous waste," for which much more stringent rules apply and permitting can take years.

## REFERENCES

[1] "Cemex, Power Play, Expensive Energy? Burn Other Stuff, One Firm Decides," John Lyons, *The Wall Street Journal*, September 1, 2004, Page A1.

[2] EPA Pet Coke report, 1999, www.epa.gov/ORD/NRMRL/pubs/600r01109/600R01109appA.pdf.

[3] ASHRAE, *Handbook of Fundamentals*, Mark S. Owen, Editor, ASHRAE, Atlanta, GA, 2005.

[4] U.S. EPA AP-42, emission factors, by chapter, fuel type, www.epa.gov/ttn/chief/ap42/ch01; for natural gas, including boilers, http://www.epa.gov/ttn/chief/ap42/ch01/final/c01s04.pdf; by industry, www.epa.gov/ttn/chief/ap42/.

CHAPTER *2*

# FUEL PROPERTIES AND COMBUSTION THEORY

## IT'S ALL ABOUT COMBUSTION

Webster's dictionary [3] defines combustion as, "a usually rapid chemical process (as oxidation) that produces heat and usually light; an act or instance of burning." In more simple terms, it involves mixing oxygen and a fuel, producing heat and products of combustion. The question is what fuel to use, and what effect will fuel switching have on emissions and performance?

Major industrial fuel users include the following:

- Boilers
- Cement kilns
- Process heaters
- Hot oil heaters
- Furnaces (for heat treating, glass making, steel, etc.)
- Dryers and calciners
- Pollution control (oxidizers)
- Waste to energy (incineration)

## COMBUSTION BASICS

The chemistry of combustion is focused on the primary reactions of oxidizing carbon and hydrogen to $CO_2$ and water vapor, respectively. The reader is referred to more detailed texts (e.g., North American Manufacturing Co.'s two-volume *Combustion Handbook* [4], and the B&W *Steam* book [5]) on the subject; the basics are covered below.

*Biomass and Alternate Fuel Systems*. Edited by McGowan, Brown, Bulpitt, Walsh
Copyright © 2009 American Institute of Chemical Engineers, Inc.

**Figure 2-1.** Some like it hot—photo of 500 MM Btu/hr furnace on a hog fuel boiler at a pulp mill.

Always keep in mind the three Ts of combustion:

- Time
- Temperature
- Turbulence

If all three exist in adequate amounts, plus the right amount of oxygen, good combustion will occur. If one or more of these are in short supply, or oxygen is lacking, bad combustion (e.g., production of soot and excessive amounts of CO) will result. Figure 2-2 shows the classic combustion temperature curve plotted versus the amount of combustion air.

Combustion engineers routinely run heat and mass balances. They can be as simple as hand calculations on the back of an envelope, but more often are either custom programs used by the engineer for a specific application, or off-the-shelf commercial programs for combustion and/or process calculations.

Examples of ready-to-use heat- and mass-balance programs are:

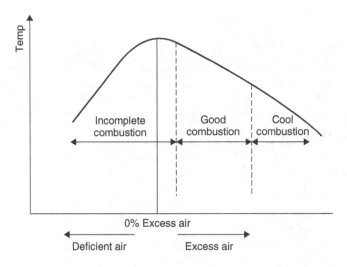

**Figure 2-2.** Temperature and combustion air.

- Hauck E-Solutions, which includes rotary dryer burner sizing, furnace sizing, emissions conversions, orifice calculations, and heat transfer. Available free via www.hauckmfg.com/solu-tions/.
- Heat and mass balance program, which includes calculations for incinerators and combustors. Free with purchase of *Intro-duction to Hazardous Waste Incineration,* 2nd Edition, Wiley, ISBN: 0-471-01790-6.
- HSC Software by Outokumpu. Performs equilibrium, heat and mass balance, and chemical formula calculations. Has thou-sands of chemicals in the database. Not for neophytes; it takes significant effort to learn how to use this software. Excellent for detailed process calculations in one or more phases. For de-scription and purchase see http://www.chemsw.com/.

Remember that mass into a system must equal the mass out. If not, you have an error in the program. Volumes (air and fuel gas) are not always equal. Figure 2-3 is a heat and mass balance for combustion of methane with 100% excess air that shows the re-lationships of mass in/mass out, transformation from reactants to products, and amount of heat liberated. Note that in the figure, the mass into the system and out of it are the same. The volume of gas (in moles, or scf) in and out is the same for methane, but

(Methane with 100% excess air)

Input

| Methane | CH$_4$ | +2 × 2 | (O$_2$ | + | 3.76N$_2$) | → | |
|---|---|---|---|---|---|---|---|
| Mols | 1 | | 2 × 2 | | 2 × 7.52 | | |
| Wt. | 16 | | 2 × 64 | | 2 × 211 | | |
| Lb/lb fuel | 1 | | 2 × 4 | | 2 × 13.2 | (Air = 34.4) | |

Out put

| | CO$_2$ | + | 2H$_2$O | + | 15.0N$_2$ | + 2O$_2$ | Totals |
|---|---|---|---|---|---|---|---|
| Mols | 1 | | 2 | | 2 × 7.53 | 2 | 20 |
| Wt. | 44 | | 36 | | 2 × 211 | 64 | 565 |
| Lb/lb fuel | 2.74 | | 2.25 | | 2 × 13.2 | 4 | 35.4 |
| Vol. %, wet | 5 | | 10 | | 75 | 10 | |
| Vol. %, dry | 5.54 | | N/A | | 83.3 | 11.1 | |
| Wt. % | 7.8 | | 6.4 | | 74.5 | 11.3 | |

**Figure 2-3.** Heat and mass balance for methane at 100% excess air.

this is not always the case. The concentration of CO$_2$ and O$_2$ can be used to back-calculate excess air levels and estimate stack gas volume. Both dry and wet basis O$_2$ and CO$_2$ concentrations are used; most test equipment produces dry basis readings. In situ oxygen CEMs are the exception, and they produce a wet basis O$_2$ concentration.

The same heat and mass balance can be run for wood or other fuels as long as the composition is known, particularly for C, H, and O. Alternately, handbooks provide mass-based ratios; for example, the stoichiometric air requirement is 765 lb per MM Btu of a particular fuel.

## Heating Value of Fuels

The heating value of fuel is the measure of heat released during the complete combustion of the fuel (with oxygen), from reactants at a given reference temperature and pressure to products at the same reference temperature and pressure.

Analytically derived formulas have been developed for prediction of the higher heating value (HHV) of coals. When compared with the wealth of knowledge generated concerning fossil fuels, interest in wood as a fuel has long been dormant. Exact calculations are available for all components of wood fuel that will oxidize; however, it is difficult to quantify the contribution of volatiles to heating value. Energy recovery from wood has stirred interest in this area, and it is expected that standards for analytical predictors of wood's higher heating value will appear in the literature.

For purposes of thermal energy generation, the higher (or gross) heating value is defined in ASTM Standard D2015-77, "Standard Test Method for Gross Calorific Value of Solid Fuel by the Adiabatic Bomb Calorimeter."

Quoting from *Steam—Its Generation and Use,* by Babcock and Wilcox:

> Most commercial fuels contain hydrogen as one of the constituents, and water, $H_2O$, is formed as a product of combustion when the hydrogen is burned in air. This water may remain in the vapor state, or it may be condensed to the liquid state, giving a substantial difference in the heat value. In determining the heat given up by a unit of such fuel, two values may be reported-the high, or gross heat value and the low, or net, heat value. For the high, or gross, heat value, it is assumed that any water vapor formed by the burning of the hydrogen constituent is all condensed and cooled to the initial temperature in the calorimeter at the start of the test. The heat of vaporization is therefore present in the reported value. For the low, or net, heat value, is assumed that none of the water vapor condenses and that all of the products of combustion remain in a gaseous state. [6]

By definition, then:

- Higher heating value (HHV) = gross heating value (GHV) is determined in the laboratory using an oxygen bomb calorimeter. HHV, expressed in units of heat released/unit weight (Btu/lb), is the heat content of a fuel (wood or other). It is always necessary, when speaking of HHV (or GHV), to note the moisture content of the sample (see Table 2.1).
- Net heating value (NHV) = lower heating value (LHV) is the net heat released by a fuel, compensating for the quantities of heat remaining in the vaporous state (as water vapor) in the flue gas.

Laboratory determination of LHV is difficult. When required, LHV may be calculated as follows:

$$Q_L = Q_H - 1040\ W$$

where:

$Q_L$ = LHV of the fuel, Btu/lb
$Q_H$ = HHV of the fuel, Btu/lb
$W$ = pound of water formed per pound of fuel combusted on a stoichiometric basis
1040 = a factor commonly used to reduce high heat value at 80°F and constant volume to low heat value at constant pressure [6].

For boiler combustion calculations, HHV is used as the basis on which fuels are bought and sold. LHVs are customarily used in Europe. When considering purchase of a foreign designed and/or manufactured equipment, the buyer should be aware of these conventions.

On an as-delivered basis, the HHVs of wood fuel will range over the values listed in Table 2-1. Note the sensitivity of HHV to moisture content. As we now can quantify, HHV is proportionally affected by fuel moisture content.

### Testing for Fuel Properties

In addition to the heating value covered above, other basic properties required for fuel combustion calculations are:

- Sulfur, nitrogen, and ash content
- Ash fusion temperature
- Major and minor constituents in the ash
- Density
- Viscosity

These properties are usually available, even for wastes. For example, waste from a Styrofoam facility would be expected to have a

**Table 2-1.** Available energy in a wood fuel at different moisture contents (higher heating value, HHV)

| Moisture content, wet basis (%) | Btu/lb | Moisture content, dry basis (%) |
|:---:|:---:|:---:|
| 0 | 8750 | 0 |
| 20 | 7000 | 25 |
| 50 | 4375 | 100 |
| 80 | 1750 | 400 |

heating value close to that of polystyrene and styrene. Properties can be found in MSDS documents for the process feedstocks, or other literature sources. When these are not available, ASTM tests can be run for heating value, proximate and ultimate analysis, major and minor constituents, moisture, ash, bulk density, and ash fusion temperature. Adding up the heating values of their individual constituents (primarily C, H, and S) can be used to approximate heating values for other fuels via the Dulong equation:

$$\text{Gross heating value in Btu/lb} = 14{,}544\ C + 62028 \times [H - O_2/8]$$
$$+ 4040\ S - 760\ Cl$$

For further information on waste and fuel testing, access the PowerPoint presentation "Matching the Process to Waste by Appropriate Testing: A Guide for Thermal Desorption Projects" on the TMTS website at http://tmtsassociates.home.mindspring.com/presentation/index.html. Similar information on testing fuels can be found on the *Chemical Engineering Magazine* website in a paper titled "Remediating Organic-laden Soils: Do Your Homework Before Breaking Ground" [1].

## BIOMASS FUELS

Energy recovery from biomass feedstocks is achieved by direct combustion or indirectly by thermochemical conversion. Direct combustion entails burning the solid biomass. Indirect methods convert the wood to a liquid or gas. The wood-derived liquid or gaseous fuel is then burned to yield heat and combustion by-products. A discussion of types of equipment and combustion processes is presented in later chapters.

## AGRICULTURAL FEEDSTOCKS

Agriculture residue (ag feedstocks, for short) are ubiquitous and tend to be dryer than wood. However, they are also seasonal and usually low in density and higher in ash, which tend to flux. The combination of lower moisture and lower ash fusion temperature is not to be taken lightly. One way around these issues is to burn them in lowered concentrations as a blend with coal or wood.

Biomass feedstocks and fuels exhibit a wide range of physical, chemical, and agricultural/process engineering properties. Despite their wide range of possible sources, biomass feedstocks are remarkably uniform in many of their fuel properties, compared with competing feedstocks such as coal or petroleum. For example, there are many kinds of coals whose gross heating value ranges from 20 to 30 GJ/tonne (gigajoules per metric tonne; 8600–12,900 Btu/lb). However, nearly all kinds of biomass feedstocks destined for combustion fall in the range of 15–19 GJ/tonne (6450–8200 Btu/lb). For most agricultural residues, the heating values are even more uniform—about 15–17 GJ/tonne (6450–7300 Btu/lb); the values for most woody materials are 18–19 GJ/tonne (7750–8200 Btu/lb). Moisture content is probably the most important determinant of heating value. Air-dried biomass typically has about 15–20% moisture, whereas the moisture content for oven-dried biomass is 0%. Moisture content is also an important characteristic of coals, varying in the range of 2–30%. However, the bulk density (and, hence, energy density) of most biomass feedstocks is generally low, even after densification—between about 10 and 40% of the bulk density of most fossil fuels—although liquid biofuels have comparable bulk densities.

Most biomass materials are easier to gasify than coal, because they are more reactive, with higher ignition stability. This characteristic also makes them easier to process thermochemically into higher-value fuels such as methanol or hydrogen. Ash content is typically lower than for most coals, and sulfur content is much lower than for many fossil fuels. Unlike coal ash, which may contain toxic metals and other trace contaminants, biomass ash may be used as a soil amendment to help replenish nutrients removed by harvest. A few biomass feedstocks stand out for their peculiar properties, such as high silicon or alkali metal contents; these may require special precautions for harvesting, processing, and combustion equipment. Note also that mineral content can vary as a function of soil type and the timing of feedstock harvest. In contrast to their fairly uniform physical properties, biomass fuels are rather heterogeneous with respect to their chemical elemental composition.

Among the liquid biomass fuels, biodiesel (vegetable oil ester) is noteworthy for its similarity to petroleum-derived diesel fuel, apart from its negligible sulfur and ash content. Bioethanol has only about 70% the heating value of petroleum distillates such as gasoline, but its sulfur and ash contents are also very low. Both of these liquid fuels have lower vapor pressure and flammability than their petroleum-based competitors, an advantage in some cases (e.g., use in confined spaces such as mines) but a disadvantage in others (e.g., engine starting at cold temperatures). Tables 2-2 to 2-4 show some "typical" values, in many cases a typical range of values, for selected compositional, chemical, and physical properties of biomass feedstocks and liquid biofuels. Note that all values are given on a dry basis, neglecting the amount of water typically found in such feedstocks at time of harvest. Figures for fossil fuels are also provided for comparison.

Further information on biomass feedstock properties can be found in the Phyllis database (www.ecn.nl/phyllis) and U.S. DOE Energy Efficiency and Renewable Energy Biomass Property Database (www.eere.energy.gov/biomass/progs/search1.cgi).

## SUSTAINABILITY OF WOOD FUELS

The question is often asked, how much wood can we grow for fuel? Sustainable growth, underutilization, unmerchantable species, and deforestation are all terms that come up in the discussion that ensues. Deforestation is a topic of great interest throughout the world, particularly in certain areas of Africa and South America (especially in the Amazon, where it is a real problem).

**Table 2-2.** Composition of certain common bioenergy feedstocks

|  | Composition | | |
| --- | --- | --- | --- |
|  | Cellulose (%) | Hemicellulose (%) | Lignin (%) |
| Corn stover | 35 | 28 | 16–21 |
| Sweet sorghum | 27 | 25 | 11 |
| Sugarcane bagasse | 32–48 | 19–24 | 23–32 |
| Hardwood | 45 | 30 | 20 |
| Softwood | 42 | 21 | 26 |
| Hybrid poplar | 42–56 | 18–25 | 21–23 |
| Bamboo | 41–49 | 24–28 | 24–26 |
| Switchgrass | 44–51 | 42–50 | 13–20 |
| Miscanthus | 44 | 24 | 17 |
| Arundo donax | 31 | 30 | 21 |

Source: U.S. DOE Oak Ridge Fact Sheet by J. Scurlock.

**Table 2-3.** Chemical dharacteristics of certain common bioenergy feedstocks and biofuels, compared with coal and oil

| | Chemical characteristics | | | | |
|---|---|---|---|---|---|
| | Heating value (gross, unless specified; GJ/MT) | Ash (%) | Sulfur (%) | Potassium (%) | Ash melting temperature (some ash sintering observed) (°C) |
| **Bioenergy feedstocks** | | | | | |
| Corn stover | 17.6 | 5.6 | | | |
| Sweet sorghum | 15.4 | 5.5 | | | |
| Sugarcane bagasse | 18.1 | 3.2–5.5 | 0.10–0.15 | 0.73–0.97 | |
| Sugarcane leaves | 17.4 | 7.7 | | | |
| Hardwood | 20.5 | 0.45 | 0.009 | 0.04 | 900 |
| Softwood | 19.6 | 0.3 | 0.01 | | |
| Hybrid poplar | 19.0 | 0.5–1.5 | 0.03 | 0.3 | 1350 |
| Bamboo | 18.5–19.4 | 0.8–2.5 | 0.03–0.05 | 0.15–0.50 | |
| Switchgrass | 18.3 | 4.5–5.8 | 0.12 | | 1016 |
| Miscanthus | 17.1–19.4 | 1.5–4.5 | 0.1 | 0.37–1.12 | 1090 (600) |
| Arundo donax | 17.1 | 5–6 | 0.07 | | |
| **Liquid biofuels** | | | | | |
| Bioethanol | 28 | | <0.01 | | N/A |
| Biodiesel | 40 | <0.02 | <0.05 | <0.0001 | N/A |
| **Fossil fuels** | | | | | |
| Coal (low rank; lignite/ subbituminous) | 15–19 | 5–20 | 1.0–3.0 | 0.02–0.3 | ~1300 |
| Coal (high rank; bituminous/ anthracite) | 27–30 | 1–10 | 0.5–1.5 | 0.06–0.15 | ~1300 |
| Oil (typical distillate) | 42–45 | 0.5–1.5 | 0.2–1.2 | | N/A |

The answer, in short, is that there is variability in the growth of biomass, with the primary limiting factors being sunlight and water, and with many others playing a part, such as temperature, elevation, soil constituents, and type of biomass.

Focusing on North America, there are several regions where the use of wood as an energy source is a viable option. In particular, areas of the upper Midwest, the Pacific Northwest, and parts of the Northeast can produce significant amount of wood fuel on a

**Table 2-4.** Physical characteristics of certain common bioenergy feedstocks and biofuels, compared with coal and oil

| | Physical characteristics | | |
|---|---|---|---|
| | Cellulose fiber length (mm) | Chopped density at harvest (kg/m³) | Baled density (compacted bales) (kg/m³) |
| Bioenergy feedstocks | | | |
| Corn stover | 1.5 | | |
| Sweet sorghum | | | |
| Sugarcane bagasse | 1.7 | 50–75 | |
| Sugarcane leaves | | 25–40 | |
| Hardwood | 1.2 | | |
| Softwood | | | |
| Hybrid poplar | 1–1.4 | 150 (chips) | |
| Bamboo | 1.5–3.2 | | |
| Switchgrass | | 108 | 105–133 |
| Miscanthus | | 70–100 | 130–150 (300) |
| Arundo donax | 1.2 | | |
| | Typical bulk densities or range given below | | |
| Liquid biofuels | | | |
| Bioethanol | N/A | N/A | 790 |
| Biodiesel | N/A | N/A | 875 |
| Fossil fuels | | | |
| Coal (low rank; lignite/subbituminous) | N/A | N/A | 700 |
| Coal (high rank; bituminous/anthracite) | N/A | N/A | 850 |
| Oil (typical distillate) | N/A | N/A | 700–900 |

sustainable basis. Large-scale wood fuel harvesting will benefit from the support of the state forestry commissions and the U.S. Forest Service to ensure that replanting of trees takes place to provide ongoing growth and sustainable yields.

The area of the United States that shows the most promise is the Southeast. It is no coincidence that it has become home to wood pellet operations and is a hotbed of developer activity for other wood fueled processes, such as cellulose to ethanol and independent power production. As a point of comparison for ongoing forestry, Figure 2-4 shows the amount of pulpwood production in the southern states. Some wood fuel processes (e.g., research at Georgia Tech on a biological route to ethanol) are based on pulpwood as the feedstock.

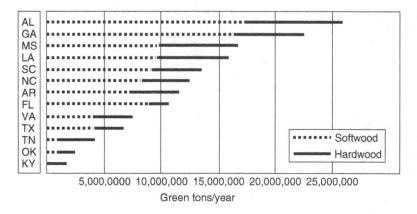

**Figure 2-4.** Pulpwood production in the southeastern United States, 2002. (Source: Georgia Forestry Commission.)

Most of this wood has been traditionally used for pulp and paper. The same forests are used for saw timber (the cycle is 10 years to thinning, 20 years to pulpwood, and 30 years for saw timber for the same stand). In recent years, however, the pulp and paper industry has been under a great deal of pressure from foreign competition and pollution laws. In fact, the total impact of forest-products-related industry just in Georgia declined from $30 billion in 2001 to under $20 billion in 2003 [2].

As can be seen in Figure 2-5, there has been some recovery of the industry since 2003, but there is continued softness in the pulp and paper industry due to foreign competition and cheaper offshore sources of pulp. This situation is typical for states in the southeastern United States; landowners are looking for new outlets for their timber and the energy markets are holding some promise.

How much more timber could be harvested on a renewable, sustainable basis? Figure 2-6 gives an estimate of the biomass available for energy use from forest land and agricultural residues. This projection includes 52 million dry tons of fuel wood harvested from forests; 145 million dry tons of residues from wood processing plants and pulp and paper mills; 47 million dry tons of urban wood residues, including construction and demolition (C&D) debris; 64 million dry tons of residues from logging and site clearing operations; and 60 million dry tons of biomass from fuel treatment operations to reduce fire hazards. This does not include

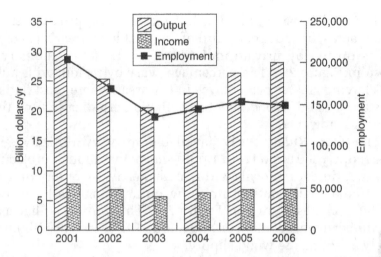

**Figure 2-5.** Economic impact of forest product industry in Georgia. (Source: W. Riall, Georgia Tech Economic Development Institute.)

windfalls from hurricanes and major storms, which, though being unpredictable, can produce significant amounts of biomass that can be used for fuel.

Not all C&D waste is fit for fuel wood. Treated wood (PCP, CCA, etc.) and painted wood (which raises lead issues) have extra environmental concerns.

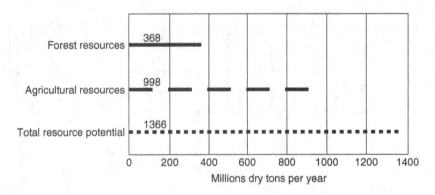

**Figure 2-6.** Annual biomass resource potential from forest and agricultural resources. (Source: Oak Ridge National Laboratory, "Biomass as Feedstock for a Bioenergy and Bioproducts Industry: The Technical Feasibility of a Billion-Ton Annual Supply," Oak Ridge, TN, 2005.)

The total forest land in the United States is approximately 750 million acres, or about one-third of the total land area. Most of this land is owned by private individuals or the forest industry, as shown in Figure 2-7. The percentages vary by region, with public lands having greater percentages in the areas where most national parks are located. Wood may be available periodically from these areas also, however.

The Oak Ridge study (available at www.ornl.gov/~web-works/cppr/y2001/rpt/123021.pdf ) shows that though projections show that timber removals will remain steady, inventory of unutilized timber will continue to grow quite dramatically over the next 50 years, as shown in Figure 2-8. This indicates that many new timber using plants could be added to demand without adversely affecting the wood supply.

The types of biomass available estimated by the Oak Ridge study are summarized in Figure 2-9. *The total adds up to 368 million tons of dry biomass available each year.* The categories are clearly shown in the figure.

One category of biomass fuel that has received a great deal of attention from the U.S. Forest Service and the various state forestry commissions is the "Fuel Treatments" category, which

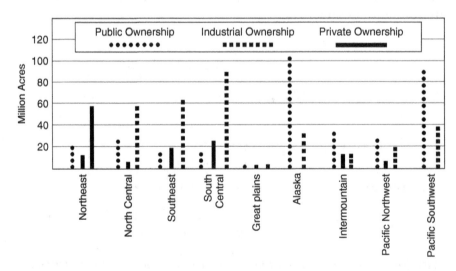

**Figure 2-7.** Ownership of U.S. forestland by region. (Source: Oak Ridge National Laboratory, "Biomass as Feedstock for a Bioenergy and Bioproducts Industry: The Technical Feasibility of a Billion-Ton Annual Supply," Oak Ridge, TN, 2005.)

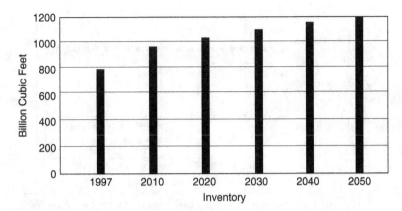

**Figure 2-8.** Projections of timber removals, growth, and inventory. (Source: Oak Ridge National Laboratory, "Biomass as Feedstock for a Bioenergy and Bioproducts Industry: The Technical Feasibility of a Billion-Ton Annual Supply," Oak Ridge, TN, 2005.)

proposes thinning of forests to help prevent forest fires. This material represents a significant class of biomass fuel, as shown in Figure 2-10.

Figure 2-11 gives a good detailed summary of the various estimated sources of wood available for energy use as determined by Oak Ridge National Laboratory and the U.S. Forest Service. It

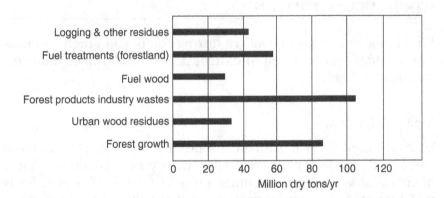

**Figure 2-9.** Estimate of sustainably recoverable forest biomass. (Source: Oak Ridge National Laboratory, "Biomass as Feedstock for a Bioenergy and Bioproducts Industry: The Technical Feasibility of a Billion-Ton Annual Supply," Oak Ridge, TN, 2005.)

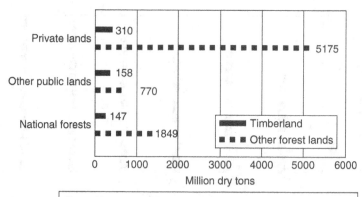

Note: About 8.4 billion dry tons treatable biomass available for biomass and biobased product

**Figure 2-10.** Total treatable biomass resource on timberlands and other forestlands. (Source: Oak Ridge National Laboratory, "Biomass as Feedstock for a Bioenergy and Bioproducts Industry: The Technical Feasibility of a Billion-Ton Annual Supply," Oak Ridge, TN, 2005.)

shows that there is a significant amount of unexplored resource, and it is apparent that that an enormous amount of biomass fuel is available from U.S. forestland that can contribute to the U.S. energy supply. Further information on the forest resource can be found at www.fs.fed.us/pnw/pubs/gtr560/gtr560_part1.pdf.

## WOOD FUEL PROPERTIES

Wood fuel properties of concern during combustion include moisture content, particle size, proximate analysis, ultimate analysis, and heating value.

### Moisture Content

Moisture content is described in one of two ways: (1) wet basis and (2) dry basis. Those concerned with power generation most often consider moisture content on a wet basis. The wet basis moisture content directly reflects the fuel value of wood. Knowledge of both methods of calculating moisture content will be important when arranging wood fuel purchases, especially mill residues.

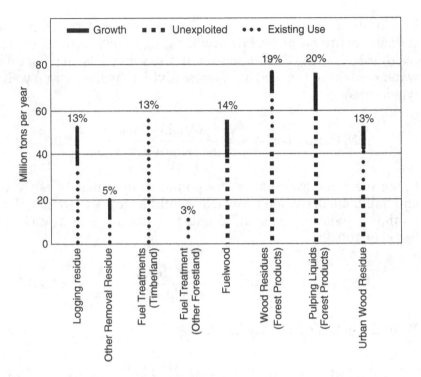

**Figure 2-11.** Summary of potentially available forest resources. (Source: Oak Ridge National Laboratory, "Biomass as Feedstock for a Bioenergy and Bio-products Industry: The Technical Feasibility of a Billion-Ton Annual Supply," Oak Ridge, TN, 2005.)

The moisture content (M.C.) of wood on the wet basis is the weight of water in a wood sample divided by the total weight of the sample:

$$\text{M.C. (wet) \%} = \frac{\text{weight of water}}{\text{total weight}} \times 100$$

EXAMPLE: A one pound sample of wood is found to have 50% M.C. (wet basis). The results of laboratory analysis showed:

$$\text{water} = \text{½ pound } H_2O$$

$$\text{wood} = \text{½ pound dry wood}$$

$$\text{M.C.} = \text{½ lb}/(\text{½} + \text{½}) \text{ lb} \times 100 = 50\% \text{ wet basis}$$

The dry-basis moisture content is favored by foresters and producer/manufacturers of wood products (a prime source of mill residues). The M.C. of wood on the dry basis is the fractional water content or the weight of water divided by the sample weight when dried:

$$\text{M.C. (dry basis) \%} = \frac{\text{weight of water}}{\text{weight of dry wood}} \times 100$$

Using the same example, a one pound sample that is half water and half bone dry wood by weight would have a wet weight of one pound, a bone dry weight of one-half pound, and a M.C. (dry basis) of 100%:

$$\text{M.C. (dry basis) \%} = \frac{\frac{1}{2} \text{ lb water}}{\frac{1}{2} \text{ weight of dry wood}} \times 100$$

To find wet basis from dry basis:

$$\text{M.C. (wet basis) \%} = \frac{\text{M.C. dry}}{\text{M.C. dry} + 100} \times 100$$

To find dry basis from wet basis:

$$\text{M.C. (dry basis) \%} = \frac{\text{M.C. wet}}{100 - \text{M.C. wet}} \times 100$$

Figure 2-12 illustrates graphically the relationship between wet basis and dry basis moisture contents. Table 2-5 covers the normally encountered range while Table 2-6 provides moisture content and heating values of typical fuels.

Unless noted, all moisture contents in this book are cited on a wet basis. High moisture content in fuelstock reduces combustion efficiency. The vaporization of water to steam requires a heat input of approximately 1000 Btu/lb of water. Energy which could otherwise be useful in steam production is thus diverted to drying the wood fuel in the combustion chamber prior to actual burning of the wood. Overall boiler capacity is also affected, as illustrated in Figure 2-13. Because the cost of equipment for predrying the fuel is high, the use of green wood fuels (M.C. between 50 to 65%) is often justifiable.

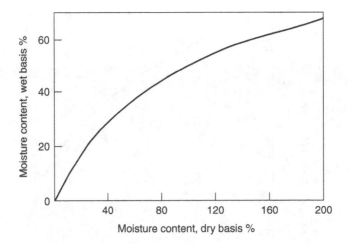

**Figure 2-12.** Moisture content comparison.

**Table 2-5.** Moisture content: comparison of wet/dry bases

| Wet basis (%) | Dry basis (%) |
|:---:|:---:|
| 0 | 0 |
| 5 | 5 |
| 10 | 11 |
| 15 | 18 |
| 20 | 25 |
| 30 | 43 |
| 40 | 67 |
| 50 | 100 |
| 60 | 150 |

**Table 2-6.** Higher heat values of some wood fuels

| Wood fuel | Moisture content, wet basis (%) | Higher heating value (Btu/lb) |
|:---|:---:|:---:|
| Whole tree chips | 50 | 4000 |
| Dry planer shavings | 13 | 6960 |
| Green sawdust | 50 | 4000 |
| Dry sawdust | 13 | 6960 |
| Wood pellets | 10 | 7200 |

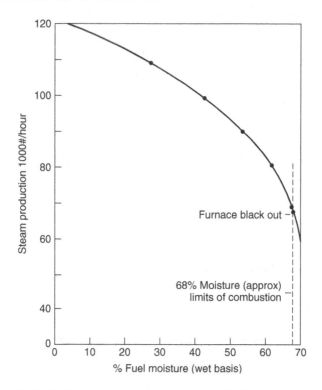

**Figure 2-13.** Boiler steam production versus fuel moisture content.

## Particle Size

The particle size of wood fuel entering the combustion chamber, more specifically, the surface area/mass ratio of the discrete pieces, will have great effect on combustion efficiency, boiler design, and quantity and method of introduction of combustion air. Improperly sized fuel may not burn completely, and valuable heat energy can be lost in the form of carbon-rich bottom ash and fly ash. Combustion units are designed specifically for certain size ranges of fuels, and close attention should be paid to maintaining equipment size requirements. Table 2-7 shows typical sizes of wood fuel (as received) for combustion.

## Proximate Analysis

Proximate analysis describes the volatiles, fixed carbon, and ash present in a fuel as a percentage of dry fuel weight. The amounts

**Table 2-7.** Typical sizes of wood fuel as received

| Type | Dimension |
|---|---|
| Whole green tree chips | 2 in. × 2 in. × ¼ in. |
| Sawdust | ⅛ in. to ¼ in. |
| Mill residue | ⅛ in. to several feet long |
| Pellets | ¼ in. diameter × ½ in. long |

of volatiles and fixed carbon directly affect the heating value of the fuel, the flame temperature, and the process by which combustion is achieved. Keep in mind that all fuels other than carbon and metals burn as a gas. As can be seen in Table 2-8, the volatile fraction of dry wood is very high, and a nominal value for wood is 80%. That means that 80% of the weight of dry wood will burn in the gas phase, with the remaining carbon burning as a solid on the grates, or as a fine particulate. The ash content is important in the design of air pollution control equipment, boiler grates, and bottom-ash-handling equipment. From Table 2-8, it can be seen that wood has far lower ash content than coal, a benchmark for solid fuel. Wood has similar higher volatiles and less fixed carbon. Ash contents in the range of 0.5 (clean stem wood) to 3% (with bark and dirt) for wood fuels have also been noted in the literature.

## Ultimate Analysis

The ultimate analysis (Table 2-9) of a fuel describes its elemental composition as a percentage of the sample's dry weight. Note that wood fuels are almost devoid of sulfur. Coupled with low ash con-

**Table 2-8.** Proximate analysis of wood and coal

| | Moisture content (%) | Volatile matter (%) | Fixed carbon (%) | Ash (%) |
|---|---|---|---|---|
| Bituminous coal | 2.5 | 37.60 | 52.90 | 7.00 |
| Hardwood (wet) | 45.6 | 48.58 | 5.52 | 0.30 |
| Hardwood (dry) | 0 | 89.31 | 10.14 | 0.56 |
| Southern pine (wet) | 52.3 | 31.50 | 15.90 | 0.29 |
| Southern pine (dry) | 0 | 66.00 | 33.40 | 0.60 |

**Table 2-9.** Ultimate analysis of wood species (percent of dry fuel weight)

| Wood species | Hydrogen | Carbon | Nitrogen | Oxygen | Sulfur | Ash |
|---|---|---|---|---|---|---|
| | | | Wood | | | |
| California redwood | 5.9 | 53.5 | 0.1 | 40.3 | trace | 0.2 |
| Western hemlock | 5.8 | 50.4 | 0.1 | 41.4 | 0.1 | 2.2 |
| Douglas fir | 6.3 | 52.3 | 0.1 | 40.5 | trace | 0.8 |
| Pine (sawdust) | 6.3 | 51.8 | 0.1 | 41.3 | trace | 0.5 |
| | | | Bark | | | |
| Western hemlock | 6.2 | 53.0 | 0.0 | 39.3 | trace | 1.5 |
| Douglas fir | 5.8 | 51.2 | 0.1 | 39.2 | trace | 3.7 |
| Loblolly pine | 5.6 | 56.3 | — | 37.7 | trace | 0.4 |
| Longleaf pine | 5.5 | 56.4 | — | 37.4 | trace | 0.7 |
| Shortleaf pine | 5.6 | 57.2 | 0.4 | 36.1 | trace | 0.7 |
| Slash pine | 5.4 | 56.2 | 0.4 | 37.3 | trace | 0.7 |

tent, these two properties make wood a highly desirable fuel from the standpoint of pollution control costs.

## PHYSICAL/CHEMICAL CONSIDERATIONS

Combustion of wood fuel occurs in several stages (Figure 2-14). The surface of the wood initially undergoes thermal breakdown into vapors, gases, and mists, some of which are combustible. The first stage of combustion exists up to 395°F. In this zone, there is a

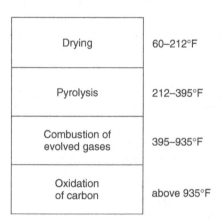

**Figure 2-14.** Stages of wood fuel combustion.

slow, steady weight loss as water vapor and other nonignitable gases are driven off.

In the temperature range of 395 to 535°F, more gases are driven off and heat-liberating reactions first occur; however, there is no flaming until higher temperatures are reached. In the next zone, temperatures range from 535 to 935°F; in this zone, gases continue to evolve and react, producing heat. At first, they are too rich in carbon dioxide and water vapor to sustain flame, but secondary reactions occur, forming combustible gases that ignite and flame. Finally, all the gases and tars are driven from the wood, and pure carbon (usually referred to as charcoal) remains. Combustion of the charcoal occurs, and the temperature of the wood surface rises above 935°F. Since the four stages of combustion occur simultaneously, many secondary reactions result, which further complicate combustion. Although all direct combustion of wood occurs in these four stages, there are sufficient differences between available biomass and wood fuels and heating applications to require different types of combustion equipment. This is dealt with in later chapters.

## REFERENCES

[1]  McGowan, T. "Remediating Organic-laden Soils: Do Your Homework before Breaking Ground," Chemical Engineering Magazine, April 2, 2004.

[2]  Riall, W. "*Economic Benefits of the Forestry Industry in Georgia: 2006*" Enterprise Innovation Institute, Georgia Institute of Technology, 2006.

[3]  *Merriam-Webster's Collegiate Dictionary,* Eleventh Edition, Springfield, MA, 2006.

[4]  *North American Combustion Handbook,* North American Manufacturing, Cleveland, Ohio, Volume 1, 3rd ed., 1986, and Volume 2, 3rd ed., 1997.

[5]  *Steam Generation and Use,* Babcock and Wilcox, Barberton, Ohio, 41st ed., 2005.

[6]  *Steam Generation and Use,* Babcock and Wilcox, Barberton, Ohio, 37th ed., 1963, page 4-3.

CHAPTER **3**

# LIQUID FUELS FROM BIOMASS

## INTRODUCTION

Liquid fuels from biomass and wood are aimed at the transportation market, with the exception of direct burning of used vegetable oil in industrial applications. Industrial use of used vegetable oil is primarily for boiler fuel at rendering plants (which produce used vegetable oil and animal fats) when fuel cost is high and when market prices for reclaimed vegetable oil or animal fats are low.

The primary biomass-based liquid fuels are biodiesel and ethanol. Current ethanol and biodiesel feedstocks are both crop based and potentially compete with food for acreage. Both have significant federal subsidies and tax incentives for motor fuel applications, which helps economics. However, both suffer from high feedstock costs, in part due to recent government initiatives to raise the rate of production, triggering feedstock price increases due to the irrevocable law of supply and demand.

Pyrolysis oil is not covered here, but rather referred to in brief in the section on gasification, and is not suitable for transportation fuel.

The following definitions are used in this chapter:

- **Cooking oil** is the material used in restaurants for deep-fat frying. It is generally vegetable oil, but some lard and animal fat may be used, and waste oil will contain some animal fat from the products being cooked.
- **Petrodiesel** means petroleum-based diesel. It is similar to No. 2 fuel oil, which is typically used for heating. Petro diesel has very low sulfur.

- **B20 biodiesel** is a 20% blend of treated vegetable oils and 80% petrodiesel.
- **B100,** also called 100% biodiesel, is 100% vegetable-based diesel fuel, not blended with petroleum products.
- **Ethanol** is blended with gasoline, and **E85** is 85% ethanol, 15% gasoline. Ethanol is blended in lesser amounts as an octane booster and oxygenate, typically up to 10%, but higher in countries such as Brazil.

### Heating Value and Properties of Liquid Biomass-Based Fuels

There are slight variances in the reporting of properties of fuels between sources. This is partly because fuels are produced to meet a certain standard, which usually has a range of acceptability. As a result, the actual properties of fuels change slightly depending on the process by which they are produced. A second reason is that the fuel is not always referred to in a specific enough manner; for example, referring to gasoline as opposed to regular unleaded gasoline, premium gasoline, reformulated gasoline, and so on.

A comprehensive and detailed description of the properties of various biomass-based and conventional fuels has been compiled by the DOE and is shown in Table 3-1.

**Table 3-1.** Properties of biomass-based and conventional liquid fuels [1]

|  | Gasoline | No. 2 Diesel | Biodiesel* | Ethanol[†] | Methanol |
|---|---|---|---|---|---|
| Octane number[‡] | 84-93 | — | — | 110 | 112 |
| Cetane number | — | 40–55 | 48-65 | — | — |
| Energy content (Btu/gal) Higher heating value | 124,340 | 137,380 | 127,960 | 84,530 | 65,200 |
| Energy content (Btu/gal) Lower heating value | 116,090 | 128,450 | 119,550 | 76,330 | 57,250 |
| Volumetric energy density (as % of gasoline, LHV basis) | 100% | 111% | 103% | 66% | 49% |

*Refers to pure biodiesel, B100. Other blends (B5, B20, etc.) have varying properties.
[†]Refers to pure ethanol, E100. Other blends (E10, E85, etc.) have varying properties.
[‡]Octane number refers to the average of the research octane number (RON) and motor octane number (MON).

# VEGETABLE OIL AND BIODIESEL

## Virgin Vegetable Oil

Virgin vegetable oil product comes from the pressing of a range of agricultural feedstocks, such as soybeans, rapeseed, cottonseed, and corn. The generic formula is C–H–O; it is high in carbon by weight and lower in hydrogen and oxygen. The oil is approximately 10% oxygen by weight, and this decreases fuel value on a mass basis by a corresponding amount.

The virgin oils have a high flash point, as expected for use in deep fat frying. Frying conditions are normally in the 160°C (300°F) to 185°C (365°F) range; hence, significant breakdown, polymerization, or decomposition is not expected if oil is kept below these temperatures.

## Used Cooking Oil

The used cooking oil hauled off from restaurants will generally be vegetable oil contaminated with animal fat (fat is defined as a solid at room temperature) and other material from cooked food. Darling International (www.darlingii.com/finished/fats_oils.htm) is one firm involved in picking up waste vegetable oil from restaurants and in processing it into finished products. Their restaurant "cooking oil" pickup service information can be found at www.darlingii.com/restaurant/corservice.htm.

In cold weather, used cooking oil may form layers in the storage tanks, with a solid waxy layer, and/or a dispersed emulsion. This solid phase will appear at the "cloud point" and will disappear on reheating. Pumping of the material under cold winter conditions without heating may plug filters with the waxy material. Duplex strainers would be a good idea, with the ability to change over while operating, allowing removal, checking, and cleaning a filter without interruption of fuel flow. Regarding the refined B100 product, the National Biodiesel Board recommends that biodiesel be kept in the 80°F to 130°F range for local shipment to prevent solidification of some constituents prior to delivery and loading. Insulated and heated tanker trucks could be used to keep waste cooking oil warm, and could be used to provide direct feed to a burner system.

The amount of tramp water picked up with the waste cooking oil will vary with housekeeping at each restaurant. In storage tanks, free water may contribute to growth of yeast, fungi, and bacteria that grow at the water-to-fuel interface. This may produce sludge and plugging, and may produce acids. Having fuel draw-offs above the bottom of the tank, and checking tanks with Tanktex or equivalent to detect the depth of the water layer (see www.rectorseal.com) and drawing off water occasionally should keep water out of the combustion system.

Vegetable oils are not the same as petroleum products, and care must be taken with wetted parts. Compatible elastomers are required, and nitrile rubber compounds, polypropylene, polyvinyl, and Tygon, and natural rubber are *incompatible*. Materials such as Teflon, Viton B, fluorinated plastics, and Nylon are compatible with B100. If seals or hoses swell, this is a sign of degradation and that they must be replaced with an appropriate material.

Viscosity is higher than that of petrodiesel/No. 2 fuel oil, and this can lead to poorer atomization of the fuel spray unless heated to the appropriate temperature. Viscosity also increases more rapidly as temperature is decreased compared to petrodiesel. Impurities also tend to significantly increase the viscosity of biodiesel. Based on Table 3-2 viscosity, and assuming the same viscosity/temp curve typical for fuel oils as an approximation only, about 30°F higher temperature is required for B100 to have viscosity equal to petrodiesel/No. 2 fuel oil.

### Manufacture of B100

B100 is upgraded vegetable oil, or virgin vegetable oil, obtained by treating it with methanol and lye. The reaction (transesterification) substitutes methanol for the glycerin in triglycerides, making methyl esters (the B100 product) plus byproduct glycerin. A sample MSDS for B100 can be found at www.biodiesel.org/pdf_files/fuel fact sheets/MSDS.pdf. Although the B100 product is not identical to used vegetable oils, many of its physical properties are similar due to the shared feedstock.

### Combustion of B100

The National Biodiesel Board (www.biodiesel.org) is a good source of information on heating value for the refined product. They show B100 having 12.5% less energy per pound as compared to petrodiesel, with B100 having 16,000 Btu/lb or 118,170 Btu/gal, versus typical petrodiesel/No. 2 fuel oil with 18,300 Btu/lb or 129,050 Btu/gal [2] (shows good agreement with Table 3-1).

Tests run on B20 with a mechanically atomized nozzle on an oil burner firing a boiler in Warwick, RI showed a slight decrease in NOx (67 ppm vs. 83 ppm) and a slight increase in thermal efficiency (79.4% vs. 78.6%). No changes were made to the firing system (a PowerPoint presentation on the Warwick project can be accessed via www.biodiesel.org/markets/hom). Note that switching a burner to B100, or a blend of biodiesel and fuel oil, or used vegetable oil, without making any other changes will automatically increase excess air levels, as hourly fuel heating value input will drop due to lower heating value per lb, whereas the combustion air rate will be constant.

The University of Georgia (UGA) ran combustion tests, burning fat and grease in a No. 2 fuel oil fired boiler. Data was taken on fuel viscosity, physical properties, and emissions. Particulates in the first load of chicken fat caused plugging problems, but this did not reoccur in the rest of the tests.

Burner manufacturers North American Manufacturing, Eclipse, and Hauck were contacted regarding combustion experience with vegetable oils. Hauck says they have experience burning waste oil, but do not recall burning vegetable oil and have no case studies on it. John Stanley at Eclipse says that there will be no problem firing vegetable oils with their Vortometric burners if solid particles are filtered out. The Vortometric burner is a swirl burner, which aids in flame retention and stabilization, an asset when burning liquid waste fuels. North American says they have experience burning vegetable oils, but no hard data was provided on past projects.

An issue for burners with mechanically atomized nozzles is preheating the fuel to reduce it to normal No. 2 fuel oil viscosity range. This may be required only during the winter. Alternately, a different nozzle could be installed to handle higher viscosity fuels. For air-atomized nozzles, the fuel must be heated to keep the viscosity at the recommended viscosity level in SSU. For either type of nozzle, the fuel must be above its cloud point to prevent formation of solids. In summary, viscosity must match the requirements of the atomizer, whether mechanically atomized or compressed air, dual-fluid type. Due to potential for plugging small ports in a mechanically atomized nozzle, the dual-fluid type would be the better choice.

**Table 3-2.** Selected properties of typical petrodiesel/No. 2 fuel oil and B100 [2]

| Parameter | Petrodiesel/No. 2 fuel oil | B100 Fuel Standard ASTM D975, ASTM D6751 |
|---|---|---|
| Lower heating value, Btu/gal | ~129,050 | ~118,170 |
| Kinematic viscosity, @ 40°C mm²/sec | 1.3–4.1 | 4.0–6.0 |
| Specific gravity kg/l @ 60°F | 0.85 | 0.88 |
| Density, lb/gal @ 15°C | 7.079 | 7.328 |
| Water and sediment, vol% | 0.05 max | 0.05 max |
| Carbon, wt % | 87 | 77 |
| Hydrogen, wt % | 13 | 12 |
| Oxygen, by dif. Wt % | 0 | 11 |
| Sulfur, wt %* | 0.05 max | 0.0 to 0.0024 |
| Boiling point, °C | 180 to 340 | 315 to 350 |
| Flash point, °C | 60 to 80 | 100 to 170 |
| Cloud point, °C | −15 to 5 | −3 to 12 |
| Pour point, °C | −35 to −15 | −15 to 10 |
| Cetane number | 40–55 | 48–65 |
| Lubricity SLBOCLE, grams | 2000–5000 | >7,000 |
| Lubricity HFRR, microns | 300–600 | <300 |

*Sulfur content for on-road fuel was lowered to 15 ppm maximum in 2006.

The literature notes that vegetable oils can clean deposits left by petrodiesel or other fuel oils in tanks and lines, which is to say, it can act like a solvent. Hence, initial changeover of tanks to vegetable oils may liberate debris [2].

For the fatty acid compositions of a number of common vegetable oils and animal fats, see [3].

## ALCOHOL FROM BIOMASS

### Ethanol from Grain

Liquid fuels from biomass have been the subject of much press and significant research in recent years. The most topical and commercialized process has been ethanol from grain, produced via fermentation. Although produced in substantial quantities in the United States and in South America, it is hindered by poor energy balance, with many studies reporting that the amount of fuel and energy required to make a gallon of ethanol is about equal to its fuel value. Ethanol from corn is currently high in cost, and in the United States significant government subsidies are required to make it commercially feasible. Finally, its oxygen content makes its heating valve significantly less than that of gasoline; approximately two-thirds, as shown in Table 3-1. Recent political initiatives to produce more ethanol have raised the price of feedstocks and stoked the food-versus-fuel debate.

## Ethanol from Wood or Cellulosic Feedstocks

One way to mitigate the cost and energy balance issues in producing ethanol is by using wood or biomass as the feedstock. If wood is used as the feedstock, the cost is decreased and the energy used to provide the feedstock is dramatically decreased. In addition, the modest reductions in net greenhouse gas emissions from corn ethanol production (~25%) are largely increased with the production of cellulosic ethanol (>75%).

If a large ethanol facility were sited and its feedstock use exceeded that of locally available waste wood, whole tree chips could be used at higher cost but with a more secure supply. Very large facilities have longer shipping distances and may dry up the low-cost resources. However, those with rail sidings may reduce this impact and allow greater shipping distances without major cost increases.

There are two major routes to produce alcohol fuels from cellulosic biomass:

1. Thermochemical—gasification followed by catalysis, commonly known as the "thermochemical platform"
2. Biochemical—breakdown into simple sugars and then fermentation by organisms, commonly known as the "sugar platform"

The U.S. government has stated that its goal is to make cellulosic ethanol competitive with gasoline by 2012. To ensure this, the U.S. DOE awarded $385 million in February 2007 to six ventures to produce ethanol from cellulosic biomass at a commercial scale. In January 2008, $114 million and then another $86 million in April 2008 were awarded to a total of seven ventures to construct demonstration scale (~1/10th commercial scale) cellulosic ethanol production facilities. In addition, several other awards were made to encourage the development of enzymes, fermenting organisms, and feedstocks for the cellulosic process.

The funding for these facilities is approximately split between thermochemical and biochemical routes. The first to break ground was in Treutlen County, Georgia. The developer is Range Fuels, Inc. The construction cost is projected to be $225 million, with ultimate production of 100 million gal/yr of ethanol/methanol and other alcohols. The feedstock is forest residues and agricultural waste. The process is thermochemical (gasification) followed by a catalyst [4].

Many other processes are being worked on. For example, Georgia Tech is working at pilot scale on a biochemical-based process using pulp wood as the feedstock to produce ethanol.

### Methanol

Methanol can be made from wood via gasification. Although the process is proven, there are no commercial facilities in existence. Methanol has similar properties to ethanol, but it is not in the mainstream of commercial motor fuels, is toxic, and acts aggressively on some synthetic fuel system materials. For these reasons, it might be best to turn methanol from biomass (or other sources, such as reforming of natural gas from remote gas wells) into synthetic gasoline by currently available processes, rather than attempt to create yet another motor fuel supply and use system.

### REFERENCES

[1]    www.eere.energy.gov/afdc/fuels/properties.html.
[2]    www.biodiesel.org/resources/reportsdatabase/reports/gen/ 20040901 _GEN-351.pdf.
[3]    www.me.iastate.edu/biodiesel/Pages/bio2.html.
[4]    *Atlanta Journal Constitution,* July 3, 2007.

# CHAPTER 4

# BIOMASS COMBUSTION EQUIPMENT—STEAM, HOT OIL, AND HOT GAS

Equipment discussed in this chapter includes:

- Wood and biomass-fired package boilers
- Equipment for retrofit of existing coal boilers to wood fuel
- Cyclone and suspension burners
- Wood gasifiers and pyrolysis units
- Fluidized-bed combustors
- Hot oil heaters
- Hot gas (hot air) heaters
- Other combustion systems

For this discussion, applications of the above are for plants having steam requirements up to 50,000 lbs/hr or equivalent hot oil or hot gas loads. Larger wood and biomass-fueled power plants generally mandate field-erected construction. Field-erected units, common in the pulp and paper industry, fall in a separate economic category.

## BOILER TYPES

For boilers, as with most large purchases of this type, the buyer should visit a working installation similar to what theirs will become. Most boiler manufacturers will be glad to arrange such a visit (a more complete list of boiler vendors is included in Appendix 1). Buying the first one of anything has its risks, and the buyer

*Biomass and Alternate Fuel Systems.* Edited by McGowan, Brown, Bulpitt, Walsh
Copyright © 2009 American Institute of Chemical Engineers, Inc.

should be clear as to whether they expect "tried and true" or are on a "R&D" track and are willing to take on more risk.

Detail on various designs of boilers, their application, and equipment offered by equipment vendors follows.

### Firetube and Watertube Boilers

Boilers may be divided into two general classes: firetube boilers and watertube boilers. In the firetube design, the heated gases travel through steel tubes passing through a water jacket; in the watertube design, the water being heated passes through steel tubes heated on the outside by the hot gases from the combustion process (Figure 4-1). There are many applications in which the firetube unit has distinct advantages over a watertube unit, particularly in light and medium industrial environments. The firetube boiler can be cheaper to purchase initially and it can be cheaper in terms of routine maintenance, particularly with regard to water treatment. Its limitations become apparent in the 20,000 to 40,000 lb/hr range, at pressures exceeding 150 psi.

Larger shell diameters require thicker plates to withstand pressure and temperature stresses. Temperature differentials in the boiler create high stresses, and these stresses, combined with the effects of precipitates and other deposits, have caused boiler explosions in the past. Because of its smaller component sizes and

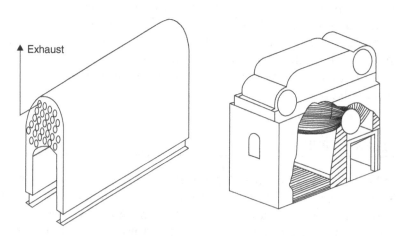

**Figure 4-1.** Firetube (left) and watertube (right) boilers.

ability to accommodate expansion, the steel watertube boiler is more suitable for large capacities and high pressure.

## Package and Field-Erected Units

Besides the firetube and watertube classifications, boiler designations can be made in another manner: boilers can be *package* boilers or *field-erected* boilers. These designations cause some confusion since virtually all wood-burning units require some field erection. A package boiler generally can be shipped by normal transportation methods such as a flat-bed truck or railcar. The major boiler components are in one assembly and can be lifted onto a simple foundation and piped into an existing system. As a result, the package boiler requires less labor before startup than a field-erected unit. The field-erected unit often requires individual welding of boiler tubes to the tube sheets or steam drum and mud drum, and the entire fabrication of a steel framework. The component parts of a field-erected boiler are completely built up at the job site, whereas the package boiler is nearly complete when it leaves the factory. Package boilers in the 200,000 lb/hr range have been shipped for gas/oil firing, but the larger combustion volumes necessary for wood or biomass units generally limit the size for a package boiler to less than 50,000 lb/hr. As one might expect, field-erected boilers cost more ($/lb/hr steam) than package boilers, and construction times are significantly longer.

## Burners and Grates

Finally, boilers may be differentiated by the method of combustion. Typical furnace/burner designs are pile burners, cyclone and suspension burners, and fluidized-bed combustors. Each system is optimally suited to particular variations of fuel inputs and types of process heat output desired.

Boilers incorporating a *pile burning design* (Figure 4-2) are used in applications in which the anticipated wood fuel has a high (up to 65%) moisture content, as is found in whole green tree chips, bark, and green mill residues. Size control of the fuel is not as critical as in a cyclone or suspension burner.

Grates serve to support the fuel while it dries and volatiles are driven off. Air flowing up through the grates (*underfire air*) serves several purposes, as follows:

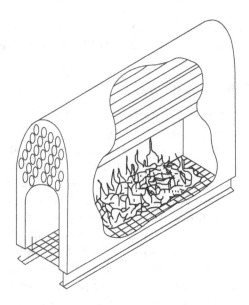

**Figure 4-2.** Pile burning on grate.

- Provides oxygen for combustion
- Cools the grates
- Promotes turbulence in the fuel bed
- Contributes to drying the fuel

The introduction of *overfire air* above the grate induces turbulence and provides oxygen for combustion of entrained and volatized material.

One type of heaped pile burner is the *Dutch oven.* Usually, fuel is gravity fed through a fuel chute onto the pile. Within the refractory-lined chamber, high temperatures are generated and the fuel is dried. Underfire air serves to partially burn the fuel and drive off the volatiles. Burning is completed in a secondary chamber where overfire air is injected (Figure 4-3).

The major advantage of a Dutch oven is its ability to utilize wet fuel of a rough, chunky consistency. A problem with Dutch ovens is their slow response to load swings, a result of the thermal inertia of the fuel pile. The fuel/air ratio changes as the fuel pile burns down, thus making control difficult. The most serious drawback to the Dutch oven is low efficiency (50 to 60%). The low efficiency results from: (1) increased heat loss due to the furnace's

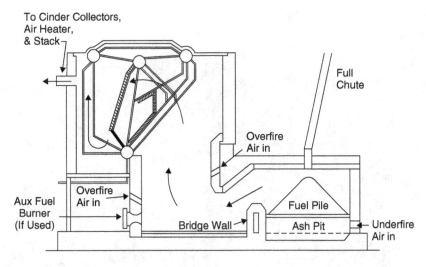

**Figure 4-3.** Dutch oven [1].

large surface area and (2) absence of radiant heating (the furnace and boiler are separate). The pile burning combustor combines well with a firetube design for steam generation, both in terms of moderate first cost and economy of size. *Turndown ratios* (the minimum percentage of the maximum design load at which a boiler will operate efficiently) may be limited, and the ability of the unit to accept coal should not be taken for granted. A Dutch oven unit is illustrated in Figure 4-3.

Gravity feed of fuel onto the pile causes two pollution problems: (1) increased probability of unburned particles being entrained and leaving the combustion zone as particulates, and (2) cooling of the combustion zone and hindrance of complete combustion. These problems can be effectively eliminated by pushing the fuel onto the pile from underneath. This is accomplished through the use of an underfed stoker.

The second class of pile burners has a thinly spread pile. Typical thin pile burners are Dutch oven or field-erected boilers with sloping grates (Figure 4-4). Fuel slides into the furnace on the grate. Particulate problems associated with dropping fuel in from above are eliminated. The slope of the grate is a function of the fuel condition. Since dry fuel slips easier than wet fuel, the grate is designed with different slopes in the drying and combustion zones. The thin pile allows more uniform air distribution as com-

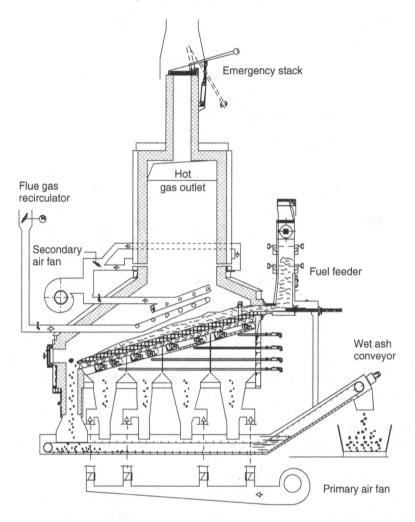

Emergency stack

Flue gas
recirculator

Hot
gas outlet

Secondary
air fan

Fuel feeder

Wet ash
conveyor

Primary air fan

**Figure 4-4.** Sloping grate furnace.

pared to a heaped pile, and combustion rates can be increased more rapidly. Wet fuel can be used, but more size uniformity is required than with a gravity-fed Dutch oven. However, there is the problem of preventing blowholes in the fuel bed, especially with nonuniform fuel. Another problem is boiler size limitations, since sloping grate boilers have not been made in sizes as large as spreader stoker boilers.

Most modern boilers (in the 20,000 to 50,000 lb/hr range) combusting coal use some type of moving grate design. Variations include *traveling grates, dump grates,* and *rotary grates* (Figure 4-5). The majority of large wood-fired boilers utilize *spreader stokers,* however. The stoker can be mechanical or pneumatic. Mechanical spreader stokers resemble a paddle wheel and "throw" fuel into the boiler, whereas pneumatic spreader stokers use air pressure to "blow" fuel into the boiler. Pneumatic stokers find wider application with wood fuel due to the size inconsistency of wood, and may incorporate rotating louvers to vary the air flow and the throw of wood particles, in an attempt to provide uniform distribution to the grate. These stokers have a high heat release rate because the smaller particles burn in suspension. The heavier particles fall to a grate where they are burned in a thin bed. Spreader stoker installations can burn wet fuel, with some drying achieved with preheated combustion air. High heat release rates, integral furnace and boiler, and lack of refractory all contribute to a smaller and lighter boiler.

In large-scale boilers (greater than 80,000 lb steam/hr), spreader stokers are used in conjunction with traveling grates to

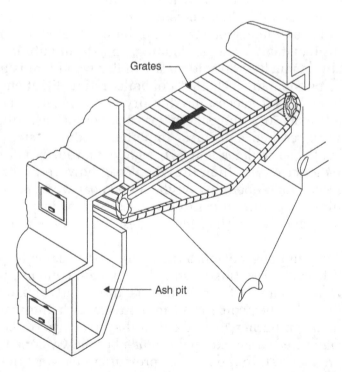

**Figure 4-5.** Traveling grate, schematic.

simplify ash removal. Babcock and Wilcox offers an enhanced stoker/furnace combustion system to lower CO and unburned carbon in biomass boilers. It uses compartmentalized grates with higher under fire air pressure, higher temperature grates, a better stoker edge seal, and better fuel distribution [2].

Disadvantages of spreader stokers are as follow:

1. Fuel interruption can extinguish the flame in the absence of refractory.
2. Overfeed leads to heavier flue gas particulate loading.
3. Heavier hogged fuel particles can fall straight down and be partially unburned when dumped since traveling grates generally rotate back to front.
4. Traveling grates have high maintenance costs.

Moving grates are generally installed on watertube steam generators; these units respond quickly to load changes and have high turn-down ratios. With minor modifications to the fuel feed systems, units designed for coal will work well with properly sized wood fuels. Should a high-moisture wood fuel (50% moisture content) such as whole green tree chips be substituted for coal, the boiler will inevitably undergo derating, and steam output may be reduced by as much as 30% to 40%. If a dry wood fuel is used, it should be possible to reach 100% of boiler rating. Plant engineers have reported good success mixing dry (15% moisture content) wood pellets 50/50 with coal. Under such conditions, full boiler ratings were reached and, most important, reduced stack emissions of both particulates and $SO_x$ were achieved. Wood pellets have a higher cost ($/million Btu) than coal; however, it has been reported that when operating on the 50/50 coal/wood pellet mixture, additional pollution control devices such as baghouses (for particulates) and FGD units (flue gas desulfurization) become unnecessary.

Cyclone and suspension burners are first cousins to pulverized coal burners. Dry (less than 15% moisture content), properly sized (less than ¼″) wood fuel is turbulently mixed with forced air and either combusted in a stream over the main fuel bed (suspension burners) or mixed in the first stage of the burner and combusted in an external cyclone burner. Cyclone burners have been successfully applied in providing heat for brick kilns in Georgia and South Carolina. Small particle-sized, dry wood fuel has high explosion potential; therefore, a flame safeguard device (such as "Fireye" by AC controls) must be provided to ensure fuel is not fed in the event of loss of flame. Boilers fired with cyclone and/or suspension burners often have high turndown ratios, good efficiencies, and excellent response to swing loads. However, they can also suffer from slag, in particular with dry fuel with higher heating value, and when using agricultural fuels with higher ash content and ash that contains fluxing compounds and elements.

Suspension burners suffer the same problems as the overhead-feed pile burners. Entrainment of particulates makes control of flyash difficult and, at high combustion rates, the residence time of fuel may be insufficient for complete combustion. Cyclone and suspension burners are illustrated in Figures 4-6 and 4-7.

Tunnel kilns for brick manufacture have also used multiport "octopus" distributors for dry fine wood. These inject the fuel pneumatically via periodic puffs through the roof, with the material burning in suspension on top of the bricks that move slowly through the kiln on carts. Typical fuel is dry hardwood shavings and dust and small chips from hardwood flooring manufacture. Some bricks contain sawdust, and as the brick is fired, the sawdust loses its volatiles, which burn in the hot kiln environment. This results in a lighter brick which has lower shipping cost per unit.

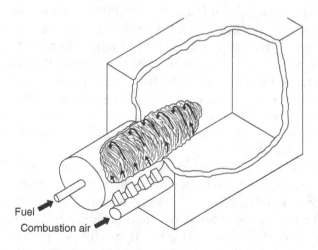

**Figure 4-6.** Cyclone burner, schematic.

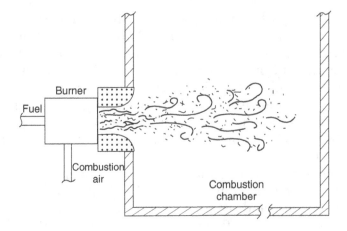

**Figure 4-7.** Suspension burner.

## WOOD-FIRED PACKAGE BOILERS—MANUFACTURERS

Wood-fired package boilers represent proven, commercially available, reliable technology. They can also fire ag fuels, as long as particle size and moisture content are in the correct range for combustor design. Many vendors are active in the wood-fired package boiler field. In general, the larger and perhaps better known boiler manufacturers (Babcock and Wilcox, Combustion Engineering, Riley Stoker, Foster-Wheeler, etc.) are not really interested in building small (less than 50,000 lb/hr) boilers for light and medium-sized commercial and industrial operations. This gap has been filled by a number of smaller manufacturers, and the market has become quite competitive. Depending on the type of wood fuel to be used, an industrial customer should be able to receive at least several bids on a complete wood burning system. The cost will be substantially higher than the traditional gas/oil package boiler, but the shorter payback time due to fuel cost savings can be attractive.

Most wood-fired boilers on the market are automated or nearly automated. A full-time boiler operator is usually required, however. System malfunctions that may occur will often be in the wood handling system. Most package boilers will require flyash collection devices to meet local air pollution codes, and the prospective industrial customer should investigate all local and state regulations before signing a contract. In addition, the contract should ensure that the boiler manufacturer or builder will guarantee compliance with the air pollution regulations.

The initial cost of wood burning package boilers can be quite high. The less expensive range of boilers are generally the ones with the least flexibility with regard to the quality (i.e., moisture content, size, etc.) of fuel that may be burned. As with suspension wood burners, even the smallest package boilers require a given amount of solids handling equipment and control systems. The system cost per pound of steam is much higher for the smaller systems. A given size of wood package boiler will generally have an initial cost of four to five times that of a comparable gas/oil boiler.

It may be difficult to retrofit an existing steam plant with a wood-fired package boiler due to space limitations. Wood systems generally require a large area for wood handling and storage, and the combustion volumes are much greater than they are for gas or oil, dictating a larger physical boiler/combustion chamber size for a given steam output.

The manufacturers and equipment listed below are presented as a representative sample for illustrative purposes only. It is by no means a comprehensive survey of available equipment. No endorsement of any particular unit, expressed or implied, is meant.

## Industrial Boiler Company/Cleaver Brooks

The Industrial Boiler Company of Thomasville, Georgia, installed over 65 wood-fired boiler systems in 23 states prior to 1984. They are now part of the Cleaver Brooks boiler company and produce gas- and oil-fired equipment. Although they no longer manufacture wood-fired equipment, the history of their primary product for the wood industry, an HRT-type (horizontal return tubular) firetube boiler (Figure 4-8) is worth exploring, and the design can be had from other vendors. Industrial Boiler made HRTs in sizes from approximately 2,600 to 34,000 lb steam/hr in single-boiler installations. (For larger steam needs, use of two or three boilers in parallel makes sense both for first cost and to allow service on one boiler while running the others. Above three boilers in parallel, the norm is to go with a single, large field-erected boiler.) The main boiler components are factory-made, then field-erected, making this a modular or semi-field-erected design. Different options are available for grate designs and furnaces. Some installations use Dutch ovens, which require more refractory work. Operating pressures range up to 300 psi, and most installations have

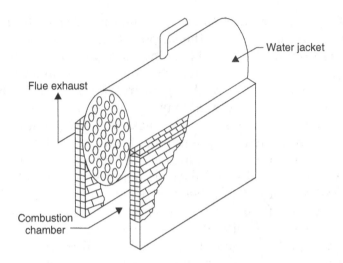

**Figure 4-8.** Horizontal-return tubular boiler.

been able to meet local pollution codes with strictly mechanical pollution control equipment, within the older guidelines of 0.08 gr/dscf (grains per dry standard cubic foot). Turnkey job installations include complete wood handling and storage systems, controls, and air pollution equipment. The manufacturer claims a thermal efficiency of 60%, and these units can burn wood waste with a moisture content up to 60%.

Advantages of the Industrial Boiler units include simplicity of design and associated low maintenance costs. Replacement of boiler tubes is a straightforward operation, and the largest single maintenance items probably will involve refractory repairs. A representative Industrial Boiler Co. installation is shown in Figure 4-9.

### Hurst Boiler and Welding Company

The Hurst Boiler and Welding Company of Cooledge, Georgia manufactures firetube and hybrid fire/watertube solid fuel combustion systems. The firetube systems can produce low-pressure steam or hot water in ranges from 3,450–20,700 lbs/hr (3.4 million Btu–20 million Btu/hr) output at 15 psig steam or 30 psig hot water. The boiler is a three-pass design with an underfed or manually fed firebox that can handle biomass fuels with moisture contents of 8 to 50%.

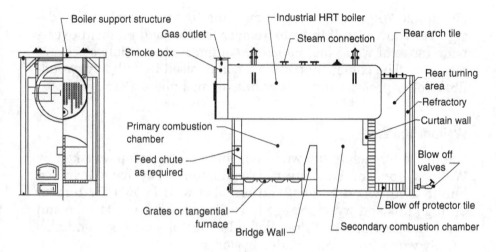

**Figure 4-9.** HRT Boiler for wood firing. (Courtesy Industrial Boiler Co.)

The hybrid systems are combinations of firetube and water-tube boilers. They can produce high-pressure steam or hot water in ranges from 3,450–60,000 lbs/hr (3.4 MM Btu/hr–60 million Btu/hr) output from 100 up to 400 psig. The system is available with a firebox with a reciprocating grate, revolving stoker bed, or flat grate bed. The system can fire biomass fuels with moisture contents of 30 to 50%, although one design is capable of firing biomass fuels with moisture contents as low as 8%.

### Teaford Company, Inc.

The Teaford Company of Alpharetta, Georgia, designs and installs field-erected boilers in sizes up to 300,000 lbs/hr of steam output. Units are available that produce saturated or superheated steam in both high- and low-pressure designs. The systems feature tangent tube, tube and tile, and membrane wall designs, and can handle both wet and dry fuels. The systems are said to have high efficiency and low environmental emissions using multistage burners.

### Biomass Combustion Systems, Inc.

Biomass Combustion Systems, Inc. of Worcester, Massachusetts, manufactures horizontal-zoned grate combustion systems in fully automated steam and hot water boiler applications from 100 to

600 hp (20,700 lb/hr steam). They claim that their combustion design has shown itself to be extremely flexible and efficient over a wide range of wood fuel moisture contents and particle sizes and densities. The combustion system can be used in both new wood-fired boiler projects and to retrofit existing boilers.

## Wellons Boilers

Another approach to the wood burning boiler concept is taken by Wellons, Inc., of Vancouver, Washington. They have installed hundreds of wood boilers, and are also well known for lumber drying kilns and wood-storage bins. They produce watertube and combination water and firetube boilers for process heat, and also supply wood-fired cogeneration systems.

Wellons uses a pile burning system that company officials call the "cycloblast furnace," which they report is capable of burning wood up to 50% in moisture content with a maximum particle size of 3″. The furnace is a refractory chamber that allows complete combustion of the fuel, but reportedly limits turndown ratios (although the manufacturer claims 5/1). Load response, due to the pile burning design, is probably slower than with other boiler configurations.

The Wellons furnace is a watertube design and is equipped with combustion air preheating. Air pollution requirements are met with a multicyclone mechanical collector (although dry ESP or baghouse controls may be required, depending on size and location) and the furnaces are operated at balanced draft with I.D. (induced draft) fan and F.D. (forced draft) fans. Figure 4-10 gives an overall view of a typical Wellons system.

## Deltak Corporation

The Deltak Corporation manufactures boilers and has supplied them in wood-fired applications. One product is called the Charger. It features shop-assembled components, developed primarily for solid fuel combustion. Its operating range is 30,000 to 110,000 lbs/hr of steam, with the ability to produce high-pressure steam that is ideal for cogeneration.

These boilers are designed for fuel to be fed by any of the conventional wood stokers, or a fluid bed combustor or gasifier/combustor. The boiler can be auxiliary fired with fossil fuel, thus al-

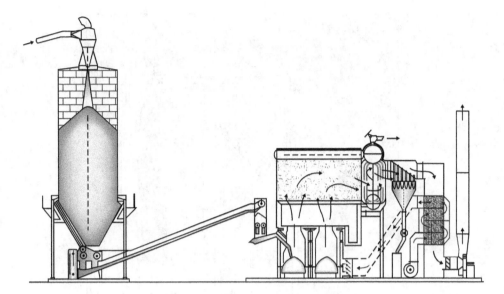

**Figure 4-10.** Wood-fired package boiler system. (Courtesy Wellons Inc.)

lowing for wet wood fuel to be burned. Deltak claims that wood waste with a moisture content of up to 55% can be burned and that boiler efficiency is in the range of 65%. However, very high wood moisture content can produce higher emissions. A typical Deltak boiler installation is shown in Figure 4-11.

## HOT OIL SYSTEMS

Thermal oil is most often used for relatively high-temperature heating applications. Most commonly, these are found in the lumber and board production industries due to the high process temperatures needed and the inherent availability of biomass from the plant's own waste streams. Thermal oil applications, however, do find their way into other industries as well.

### Thermal Oil Versus Steam—Reasons for Using Thermal Oil

Steam or thermal oil may be used to provide process heat, to some extent interchangeably, up to certain temperature ranges. However, as process temperature requirements go above approximately 450°F, steam pressures become very high. Consequently, steam

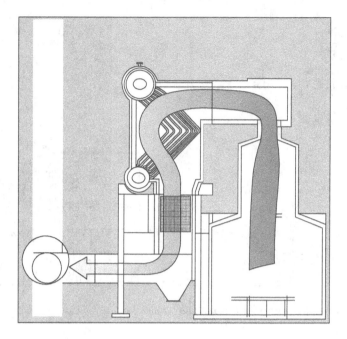

**Figure 4-11.** Modular wood-fired package boiler. (Courtesy Deltak Corp.)

system components become more expensive, and process equipment design requirements to safely operate steam systems at such high pressures become complex.

By comparison, for most mineral oil fluids, conventional thermal oil systems operate in the liquid phase at operating temperatures of up to 600°F. Most systems are nonpressurized, meaning that they are vented to the atmosphere, and operate at only normal pumping pressures required to meet the circulation needs of the plant. At 600°F, single-phase thermal oil systems operate at pressures generally less than 150 psig. In comparison, a saturated steam system at 600°F must operate at slightly over 1500 psig. Figure 4-12 shows the trend of saturated steam temperature versus pressure.

This makes the design of piping systems and components much simpler and less costly for thermal oil systems when higher process temperatures are needed. Due to the lower pressures, and the fact that thermal oil systems operate in a single phase, piping and process components may be designed to the simpler ASME Section VIII code rather than the more involved Section I power boiler code. Similarly, for single-phase systems, piping is designed to the simpler ASME B31.3 code for process piping.

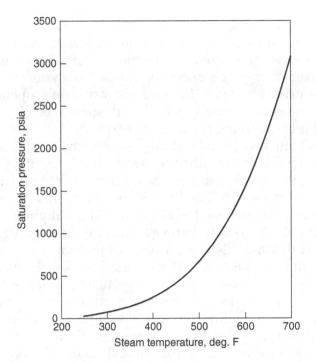

**Figure 4-12.** Saturated steam temperature versus pressure.

Some vapor-phase thermal oil systems are in use as well today that resemble steam systems to some extent in that the system operates as a two-phase system. In this case, the heat transfer fluid is heated above its boiling point and used for heating in the vapor phase, returning to the heater, or boiler in the condensed liquid phase. However, these systems are much less common, and require special safety precautions. As they are not common in biomass applications, these are not discussed further in this section.

### Types of Thermal Heat Transfer Fluids

Two kinds of heat transfer fluids are commonly available:

- Organic-based mineral oils
- Synthetic heat transfer fluids

Of these, by far the most common are the petroleum-based organic mineral oils, due to their ready availability, low toxicity, and easy handling, lower viscosity (at low temperatures), and much lower cost when compared to synthetic fluids. Mineral oils also exhibit very low vapor pressure and, therefore, they remain in the liquid phase over their entire operating range. This allows their use in nonpressurized systems. Mineral oils can safely operate in temperature ranges of 200°F to 600°F, with some specially formulated mineral oils extending the range to 650°F.

Synthetic heat transfer fluids generally consist of aromatic organic compounds or silicone-based fluids. Synthetics exhibit much higher vapor pressures than mineral-based oils and, hence, systems must be specifically designed to operate as pressurized systems in order to keep the synthetics in the liquid phase. These fluids are also generally harder to handle than standard petroleum-based thermal oils since aromatics present a personnel safety hazard, and special precautions must be taken in handling and storage. Silicone-based fluids, in particular, often crystallize at ambient temperatures, so measures must be taken to prevent this "freezing" when cooling systems or components to ambient temperatures, such as during shutdowns or maintenance work. For these reasons, synthetic fluids generally are selected only when operating temperatures exceed 600°F, when standard thermal oils cannot be used.

### Thermal Oil Properties

Several properties of heat transfer fluids must be taken into account for fluid handling and heat transfer.

Fluid density is important and has a direct effect on the pumping power required for moving the fluid through the heating circuit. Since the density will vary significantly over the range of temperatures that the fluid encounters, pump motor sizing and thermal expansion of the fluid must be accounted for over the entire temperature range. A common practice is to size pump motors for the start-up conditions based on a cold fluid. This ensures that motors do not overload on cold start. Thermal expansion must take into account the difference in oil density from cold start to operating temperatures. Typically, mineral oils will see thermal expansion of 20–25% over their operating range, with densities varying from 0.8–0.85 S.G. at ambient temperatures to 0.65–0.70 S.G. at operating temperatures of 500–550°F. Synthetic heat transfer fluid expansion is even greater, varying 50–75% over their operating ranges.

## Properties Affecting Heat Transfer

Fluid properties that affect heat transfer are viscosity, thermal conductivity, specific heat, and density. All of these must be taken into account when evaluating a thermal fluid. Thermal conductivity has a direct relation to the heat transfer coefficient and, therefore, is a strong indicator of heat transfer capability. Viscosity has a strong influence not just on pressure drop but also on heat transfer rate. As this property varies widely over the fluid's temperature range, it must be evaluated at the operating temperature. Specific heat indicates the heat carrying capacity of the fluid. The higher the specific heat, the lower the temperature differential the fluid will experience for a given heat load. This results in a higher log mean temperature differential between the fluid and the load and, therefore, higher heat transfer rates. Since all of the above properties affect the heat transfer capability of a particular oil, the best method of determining heat transfer capacity of any particular thermal fluid is to evaluate the heat transfer coefficient at the intended operating temperature of the fluid.

## Types of Thermal Oil Heaters

Thermal oil heaters come in several different arrangements, depending on the size of the heater and the specific application. In biomass applications, the simplest and perhaps lowest-cost design consists of the simple helical coil. Steel pipe is rolled into a helical coil pattern, with thermal oil on the inside and hot combustion products flowing through the center of the coil. Heat is transferred from the gases principally by radiation but also partially by convection.

Multiple coils may be combined in a double helical coil pattern, with one smaller coil placed inside the other, to form a "multipass" heater. In this design, the hot gases travel through the center coil, transferring heat principally by radiation. After passing through the center, the gases then turn 180 degrees and pass through the annulus between the inner and outer coils for a second pass. Heat transfer in this section is mostly by convection, as gas velocity between the coils is relatively high, enhancing convective heat transfer. Gas flow may then be arranged to turn around once again, passing over the outside of the outer coil to form a third pass, maximizing the overall heat transfer efficiency of this design. A three-pass dual helical coil design is shown in Figure 4-13.

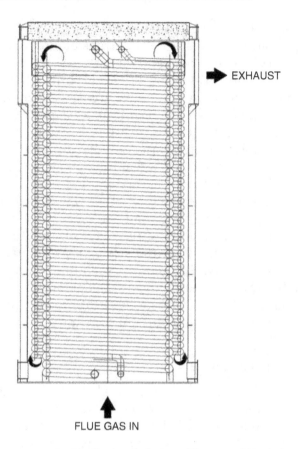

**Figure 4-13.** Three-pass dual helical coil design. (Courtesy Wechsler Engineering and Consulting, Inc.)

The advantages of the helical coil design include low cost and simplicity. However, this design does not lend itself to easy or automated cleaning of any ash deposits that may occur and obstruct flow. Therefore, care has to be taken to limit application of this type of heater to fuels with low ash content, or those that have a low fouling tendency. Even then, periodic manual cleaning by air lance may be necessary. Practical size limits for this type of heater in biomass applications is limited to usually less than 30 million Btu/hr of thermal oil output. Beyond this size, annular passages become long and access is too difficult for efficient cleaning.

Another common design is the "convective" tube bundle design. In this design, the heater tubes are arranged in a serpentine flow pattern, with the thermal oil flowing inside parallel banks of tubes, oriented perpendicular to the gas flow. The thermal oil flows down one tube, then flow is turned around via a 180° bend and the flow returns back down the next row of tubes. This type of serpentine flow pattern continues for several passes, until the oil finally exits the convective tube bundle. Heat transfer is primarily by convection, as the name implies.

This design has the advantages of high heat transfer efficiency and very large heat transfer surface area in a relatively small space. In addition, the design lends itself to automatic cleaning by soot blowing, since soot blowers can easily access the space between tubes. Generally, gas flow lanes with wide enough spacing to accommodate the soot blowers can easily be built into most convective tube bundles, or between tube bundle sections. For this reason, this type of heater design is very suitable for use in biomass applications, in which ash carryover rates can be quite high. However, the cost of construction, when compared to the helical coil design, is much higher. Convective heaters may range in size from 20 million Btu/hr up to 100 million Btu/hr of thermal oil energy output.

Many designs employ a combination of helical and convective coils. In this design, hot gases from the biomass furnace enter a single helical coil section, also called the radiative section. As the term implies, the main mode of heat transfer in this section is by radiative heat transfer. This section is then followed by the convective tube bundles, in which the gas flows through the convective tube bundles. The principal advantage of this design is that a significant portion of heat is transferred in the radiation section, cooling the gases and entrained ash to below the ash softening point, before the ash enters the convective tube bundles. This minimizes the impaction of soft, sticky ash particles onto tube surfaces, and consequent fouling of the tubes. This type of design is perhaps the most common design for larger biomass thermal oil heaters, due to the combination of efficiency, cleanability, and lower fouling potential offered compared to the single-type design.

## Thermal Oil Circulation System Components

The thermal oil circulation consists of basically two main types of circulation loops:

1. Primary thermal oil circulation loops
2. Secondary circulation loops associated with the heat users

A simplified thermal oil system is shown in Figure 4-14.

### Primary Circulation Loop

The primary loop's purpose is to maintain a safe rate of flow of the thermal fluid through the thermal oil heater, as well as supplying a sufficient flow of hot thermal fluid to all of the secondary loop heat users. For any thermal oil system, the design of this loop is critically important to the safety, reliability, and life of the thermal oil heater and associated system components. A simplified schematic of a primary loop is shown in Figure 4-15.

    In a properly designed system, the functions of the primary circulation system are to:

1. Provide adequate primary flow of thermal fluid through the heater
2. Take care of thermal expansion of the fluid

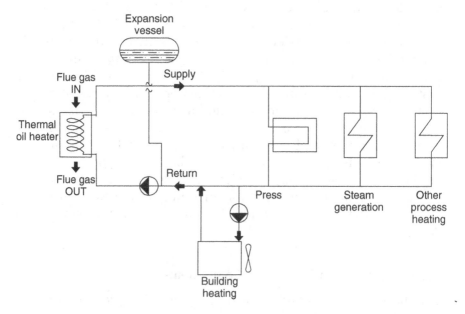

**Figure 4-14.** Simplified thermal oil system.

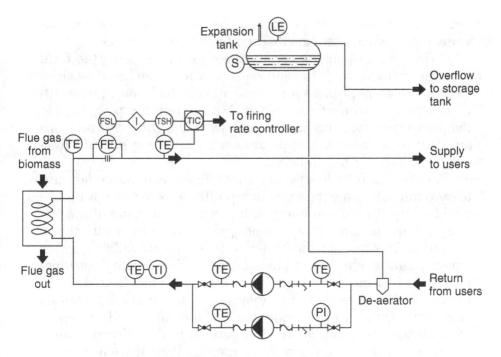

**Figure 4-15.** Typical primary loop configuration.

3. Provide means for deaerating the fluid
4. Provide the secondary loop with adequate thermal oil and, hence, heat energy supply

   Maintaining the required design flow through the primary loop and especially through the thermal oil heater is of most critical importance for any thermal fluid system. Adequate flow of thermal fluid must be maintained at all times to keep the film temperatures of the fluid in the heater coils below the heater's design value and below the maximum temperature rating of the fluid. Failure to maintain the minimum flow, or inadequately designed primary flow systems will degrade the thermal fluid. As the fluid decomposes, it forms carbon and sludges, which can build up on the inside of the tubes. This inhibits the heat transfer between tube surface, heated by the hot gases, and the thermal fluid. This, in turn, can lead to local overheating of the tube wall and reduced life of heater coils, and even eventual coil failure. For this reason, all thermal oil systems should be equipped with a flow sensing device tied to an interlock, which ensures firing of the biomass system only when adequate thermal fluid flow is proven.

Primary oil pumps are responsible for maintaining the fluid flow, generally through the entire primary loop, and must be sized to take into account pressure drops through the heater, valves, fittings, and the entire primary oil supply and return piping, back to the pump suction. If necessary, such as when primary pipe loops are very long, booster pumps are used on the supply end (discharge) of the thermal oil heater.

Most often, primary systems are designed with redundant primary pumps, to allow taking a pump offline for service or to provide backup should one pump fail. Often, one or more of the primary pumps is connected to emergency power to ensure that a minimum flow through the heater coils is always available during a power outage, since most biomass heaters have enough residual heat stored in the refractory lined sections to overheat stagnant thermal fluid, even when the biomass combustor is idled after an emergency shutdown. In some cases, a separate small, independent, emergency pump is piped in parallel to the primary pumps to provide the minimum needed emergency flow, thereby necessitating a smaller emergency power supply. This emergency circuit is tied to either a reliable thermal load that operates during a power outage, or a dedicated emergency cooler to dissipate the residual heat.

Expansion of thermal oil is normally taken up through the combination of the expansion tank, through which the fluid expands, and the drain or overflow tank, to which excess expanded fluid drains.

Thermal oil as it is heated, even when "cured" or "cooked out," will emit some organic vapor as the oil decomposes over time. These vapors need to be continuously vented from the liquid; otherwise, pump cavitation and other issues may be encountered. The deaerator provides a means by which these vapors are allowed to vent through the expansion system. This prevents buildup of gases in the thermal fluid, which could lead to pump cavitation and impeller wear or even damage.

## Secondary Loops

Generally, heat demands in thermal oil systems are supplied by the secondary loop circulation system if the primary pump pressure is not sufficient to supply a single user or if multiple users are tied to a primary loop. Typical secondary loops consist of a circulating pump, shutoff valves, and one or more control valves that control the amount of hot oil supply that is fed into the user. A typical configuration is shown in Figure 4-16; however, configurations of secondary loops vary, depending on the temperature or flow required by the user. Unlike primary loops, unless the process is very critical, redundant pumps are generally not used, since there is no heat generating equipment contained in the secondary loop that could result in oil degradation, equipment damage, or unsafe conditions.

## PRODUCTION OF HOT GAS FROM BIOMASS

Hot gas produced by direct combustion of biomass is effective for drying wood products or other bulk materials. Similar to thermal oil energy recovery, energy recovered as direct hot gas is also most prevalent in the wood-based industries due to availability of in-house generated biomass waste fuels. As the cost of fossil fuel processes rises, biomass-derived direct hot gas energy use is finding its way into other industry segments as well, including:

- Drying of raw materials (flake, wood particles) prior to pressing board in the wood board industry
- Heating of wood kilns in sawmills
- Drying of raw materials used for making fuel pellets

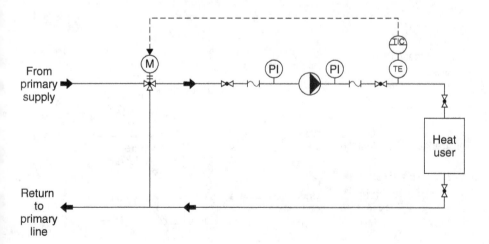

**Figure 4-16.** Typical secondary loop configuration.

## Components of Hot Gas Generation

The components of a hot gas generation system generally consist of:

- Furnace/combustion system
- Postcombustion/secondary chamber
- Hot gas control system

A schematic of a hot gas system is shown in Figure 4-17.

The furnace or combustor is generally of the type as described previously for biomass combustion. In a well-designed system, an adequate post or "secondary" combustion chamber is provided to complete combustion, and also to reduce the spark carryover into the heating process. For critical hot gas processes, additional components to reduce spark and ash carryover may be included if the downstream processes or products could be affected.

Most hot gas systems must provide hot has at a constant regulated temperature. As the quality of biomass fuel can vary significantly, and often the hot gas temperature requirement is lower than that of the combustion process, hot gas temperature regulation is generally needed. This is normally done using an air dilution or blending system, where dilution air is added to the exhaust stream of the combustor in the proper regulated amounts needed to maintain a constant hot gas supply temperature (Figure 4-18).

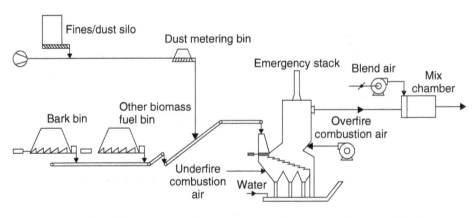

**Figure 4-17.** Schematic of a hot gas generation system for wood drying.

## COMBINED SYSTEMS: HOT GAS AND THERMAL OIL

Often, thermal oil, steam, and hot gas systems are combined into a single system when different process needs are required. In board plants, it is common practice for biomass systems to provide thermal oil process heating for various process needs, such as pressing board, steam generation, building heating, and other uses while simultaneously providing hot gas energy for the drying of wood flakes prior to the pressing of board. In some cases, this may also be combined with direct or indirect steam generation for either process needs or cogeneration of power. An example of a combined system is shown in Figure 4-19.

## FLUIDIZED-BED COMBUSTORS

The fluidized-bed combustor (Figure 4-20) has enjoyed much publicity over the past several years, particularly as an alternate combustion system for coal burning. Basically, a fluidized-bed system

**Figure 4-18.** Hot gas system supplying a wood dryer. (Courtesy of Enplant Engineering, SA.)

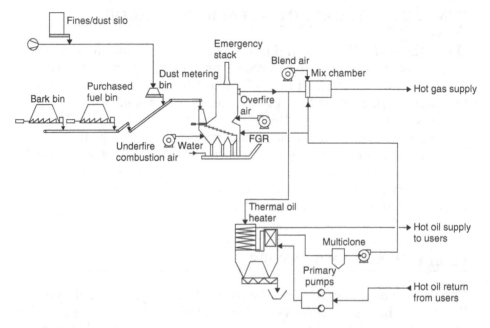

**Figure 4-19.** Schematic of combined hot gas generation and thermal oil system.

relies on a combustion chamber that has many holes drilled through a distributor plate in the floor (or pipe-type distributors), through which underfire air passes. The "bed" consists of small particles of sand, limestone, or other solid material. The bed is kept in suspension by forced air and is heated initially by an aux-

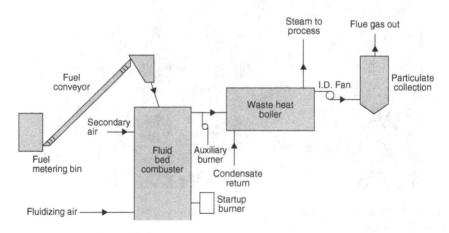

**Figure 4-20.** Fluidized-bed wood combustion system.

iliary fuel for start-up. When the bed reaches a temperature sufficiently high to ignite the fuel to be burned, the auxiliary fuel can be shut off and the solid fuel introduced. The turbulent mixing action of the hot bed material helps ensure that the fuel is burned completely. Fluidized beds have been used for coal burning because the bed material can be limestone. The limestone reacts with the sulfur in the coal to form a salt that can be removed from the bed, thus preventing the escape of sulfur compounds from the stack.

Fluidized-bed combustors can be used to generate hot gas for drying, or to supply hot gas for steam or hot oil production. In wood burning systems, the fluidized-bed combustors have shown promise as devices capable of burning wet fuels or fuels of irregular sizes and shapes. In addition, other waste materials (carpet wastes, peanut shells, etc.) may be burned in conjunction with the wood or ag feedstocks. As noted throughout this book, ag feedstocks may produce ash and be more prone to slagging than wood, and this can present severe operating problems for fluid beds.

Several companies are actively marketing various fluidizedbed systems and many units have been operated successfully in the forest products industry. Fluidized-bed systems generally have a slightly higher (perhaps 10%) first cost and higher power requirements for the fans when compared to conventional systems. These disadvantages, however, are offset by the following:

- Short residence time (rapid combustion) in the fluidized bed is conducive to a faster reaction to changes in heat demand (load swings); there is rarely more than 30 seconds of fuel in the combustion zone at any time.
- Combustion efficiency is significantly higher in the fluidized bed, resulting in considerably less unburned hydrocarbons remaining in the ash.
- High ash content fuels are readily handled in the fluidized bed. Wide variations in moisture content, sizes, and heating values of fuel are easily accommodated. Several dissimilar fuels may be fed in series or simultaneously.
- Continuous operation is achieved with the fluidized bed. There is no need to take the unit offline to remove ash and debris. An optional bed cleaner and continuous or periodic bed drains are used with dirty fuels.
- There is reduced maintenance in the combustion chamber of the fluidized bed, with simpler grates or air distributors, and

less need for cleaning, repair, or replacement. Control of combustion temperatures under 2000°F is conducive to longer refractory life and generally eliminates slagging of ash.

### York-Shipley (Division of AESYS Technologies, LLC)

A wide range of standard models are available from 5–120 million Btu/hr, as well as custom designs for unusual applications. Boiler conversions, complete steam generating systems in a variety of pressures and capacities, indirect hot-air systems with gas-to-air heat exchangers, and direct hot-gas systems for kilns and dryers are operating throughout the United States and Canada. A York-Shipley unit is shown in Figure 4-21.

### Johnston Fluid Fire

The Johnston Boiler Company of Ferrysburg, Michigan, has obtained a patent license from Combustion Systems Limited in the United Kingdom to sell in the United States fluidized-bed techniques developed in that country. The emphasis in Britain has been on coal firing using limestone as the bed medium, but the company has indicated successful operation on dry wood waste using sand as the bed medium.

The Fluid Fire unit is available in sizes from 5000 to 25,000 lb/hr of steam at pressures up to 300 psi. The Fluid Fire is a shop-assembled package-type unit transportable by rail (and by truck in the smaller sizes). As many components as possible are mounted on the unit prior to shipment to minimize field erection. Gas and oil auxiliary firing is normally provided and is nec-

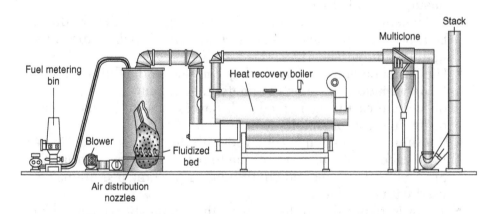

**Figure 4-21.** Fluidized-bed wood combustor. (Courtesy York-Shipley Inc.)

essary for start-up. A multiclone collector is included in the package for air pollution control. A typical Johnston unit is shown in Figure 4-22.

In addition, Johnston Boiler produces three- and four-pass water-back firetube boilers with efficiencies up to 90%. These boilers range in size from 1700 to 85,000 lbs/hr.

### Energy Products of Idaho (EPI)

EPI fluidized-bed combustors and boilers are available that can be fueled by a variety of forest, agricultural, and municipal solid waste feedstocks. EPI fluidized beds use a bed recycle system that offers uniform bed drawdown, integrated air cooling, and automatic cleaning of the bed material, and then reinjection of the bed material. This allows the system to operate on fuels with significant quantities of noncombustible tramp material such as rocks and metal scraps. To date, EPI fluidized beds have been fueled by over 200 varieties of biomass feedstocks.

It may be expected that the fluidized beds will require higher maintenance than conventional wood burners and that more high-

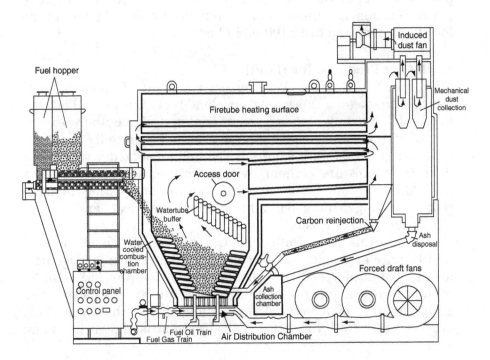

**Figure 4-22.** Package fluidized-bed boiler. (Courtesy Johnston Co.)

ly skilled operators may be necessary to operate them. Research and development in fluidized-bed technology is very active at present. Other companies working on fluidized-bed technology are included in Appendix 1.

## BURNERS AND FURNACES

### Suspension and Cyclone Burners

The cyclone furnace for burning pulverized coal has enjoyed widespread use on utility boilers for many years. The fuel is very finely ground and blown into the furnace almost as a gas, and, as a result, the combustion process is complete and efficient, and fly ash removal is easily dealt with.

Variations on the cyclone furnace concept have been developed for burning wood waste, but certain limitations can hamper the feasibility of using these systems in many applications. The wood residue must be dry (less than 15% moisture content on a wet basis), and it must be hammermilled or hogged into fairly fine particles. In spite of these requirements, several companies have placed a number of these units in industrial plants, largely in the forest products and brick (kilns) industries.

### Earth Care Products, Inc (ECPI)

ECPI (formerly Guaranty Performance Company of Independence, Kansas) markets a horizontal cyclone-type burner called the ROEMMC System which is conventional in design with the exception of an integral ash separation device that reportedly results in a cleaner gas at the unit's output. The fuel must be dry wood (less than 15% moisture content) with a uniformly small size. The manufacturer reports that the burner can run on a variety of materials including bark, peanut hulls, mesquite, peat, and several other categories of waste.

Due to the air pollution control equipment, the ROEMMC unit is quite large and may be subject to higher than average heat losses. The burner system is available in sizes from less than 10 million Btu/hr to greater than 60 million Btu/hr. ECPI also offers a similar unit in a vertical orientation, called the CycloClean burner. An overall view of a typical system is shown in Figure 4-23.

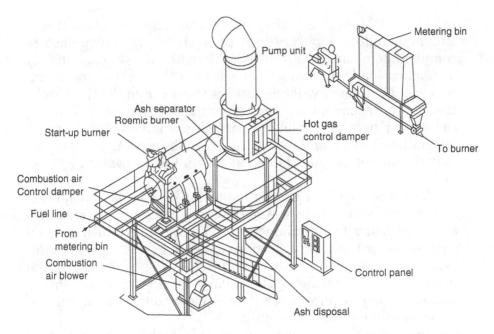

**Figure 4-23.** ROEMMC wood burner. (Courtesy Guaranty Performance Co.)

## Coen Company

The Coen Company of Woodland, California, is perhaps best known for its oil burning equipment and its low-Btu gas burners, but it has been active in wood waste systems for some years and reports that it currently has more than 5000 systems installed running on low-BTU gas and biomass. The editor of this handbook has direct experince with their burners firing sander dust in a 500 million Btu/hr pulp mill boiler.

The Coen burner is called the DAZ or Dual Air Zone and, as with previous systems, it requires dry wood waste (less than 12% moisture content) for proper operation. The size of the particles that can be tolerated depends greatly on the application of the system, as larger particles can be burned where longer residence times are available. The Coen burner consists of two air registers with concentric louvers that divide the incoming airstream into two counterrotating concentric streams. These streams provide a turbulent mixing action, which results in a compact flame pattern.

Coen reports that the DAZ burner has been fitted to incinerators, rotary dryers (for wood chips), veneer dryers, lumber drying kilns, field-erected boilers, and package boilers. Many installations include a refractory furnace or air heater, which ensures complete combustion of the fuel. As with the systems already mentioned, periodic cleanout of ash from the combustion chambers will be necessary. Particulate emissions from the Coen units are highly dependent on fuel ash content. Coen reports that it can supply burners capable of providing 5 million Btu/hr up to 100 million Btu/hr, and will do turnkey jobs as well as supplying only the burner. A typical Coen installation is shown in Figure 4-24.

The above systems give an indication of what is available for burning dry wood and ag waste. However, other systems are available, and many forest products plants build much of their own equipment for handling dry wood waste (or have it custom built). Prices for equipment vary widely due to the site specifics of the installation. The smaller systems have a higher cost per energy unit since they require roughly the same amounts of control and wood handling as the larger units.

In summary, the suspension and cyclone burners can be a viable alternative for the plant that already has a ready supply of dry wood waste available that could be reduced to a small size with a minimum of effort. Other manufacturers of this type of equipment are included in Appendix 1.

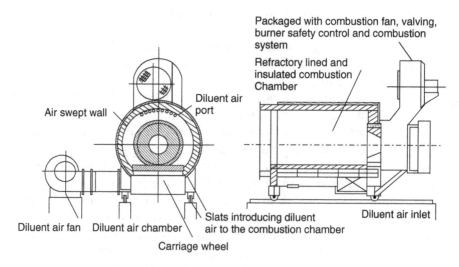

**Figure 4-24.** Dry fuel burner. (Courtesy Coen Co.)

# FURNACES FOR BURNING BIOMASS

At the heart of the biomass energy recovery system is the furnace or combustor. The role of the combustor is to reliably burn the biomass, or as it is commonly referred to in this case, fuel, in a clean and efficient manner, releasing the energy contained in the fuel. Several types of furnaces used in industrial applications today include:

- Reciprocating grate
- Stoker/pile burners
- Fluidized Beds
- Moving grates
- Suspension burners
- Other specialized combustors (e.g., rotary kilns)

Operating characteristics of the various types of several types of biomass furnaces are shown in Table 4-1.

Stokers or pile burners, as they are sometimes called, consist essentially of a fixed hearth in which the wood material is "piled" by an underfeed screw or similar device. Primary combustion air is forced through the pile while secondary air is fed over the fuel pile to help complete combustion. Ash that builds up in the pile burners is periodically removed by manual means. Although these types of combustors are still in use, they have generally fallen out of favor due to their poor emission and combustion characteristics, and high operating and maintenance requirements.

Bubbling fluidized beds are found in biomass applications; however, they have found limited use in the wood board industry, even though they are commonly found in other industrial applications. This is due to the high electrical power requirements needed to drive the fluidizing blowers and the limited range of fuel size that they are consistently able to handle. However, fluidized-bed combustors offer the highest combustion efficiency, very low emissions, and can burn very high moisture content materials.

Perhaps the most widely used type of furnace for bark and wood combustion is the reciprocating or sloping grate (Figure 4-25). Fuel is fed into the furnace directly onto the reciprocating grates via a hydraulic ram. The reciprocating grate consists of a series of alternating rows of stationary and moving bars that move in a forward and backward or "reciprocating" action. The reciprocating action of the grates continuously moves the fuel through the

**Table 4-1.** Comparison of some of the operating characteristics of biomass combustion furnaces

| Type of furnace | Fuel moisture (% wet basis) | Fuel type | Fuel size | Emissions characteristics |
|---|---|---|---|---|
| Reciprocating grate | 30–55% | Wood chips and bark, recycled wood | Fines to 4" or less, very tolerant to oversize and out-of-shape material | Low CO,* moderate $NO_x$, moderate particulate |
| Stoker/pile burners | 25–45% | Wood chips, fines and shavings, some bark | Fine material, typically 1" minus. | High CO, moderate-to-high $NO_x$, moderate-to-high particulate |
| Fluidized beds (bubbling) | 40–60+% | Wood chips, sawdust, bark | Screened size 1–3" or less. Not very tolerant to oversize or out-of-shape materials | Very low CO, low $NO_x$, high particulate[†] |
| Suspension burners | 0–10% | Sander dust or fine shavings from trimming/sawing operations on dry fuel | <1/8" particle size typical | High CO, high $NO_x$ |

*With suitably sized secondary combustion chamber.
[†]Usually requires coarse particulates and removal such as hot cyclone following the furnace stage.

furnace to ash discharge as it is burned; the ash is finally discharged to an ash conveyor.

Underfire air is fed from under the grate through the fuel for stable and controllable combustion. Secondary air and, sometimes, recirculated flue gases, is injected over the fuel bed to provide excess air and turbulence for mixing and complete combustion.

Although the reciprocating grate is perhaps not as efficient a combustor as the fluidized bed, it has found very wide use due to its forgiving nature. Some of the principal advantages of this type of furnace are its ability to handle a wide fuel moisture range,

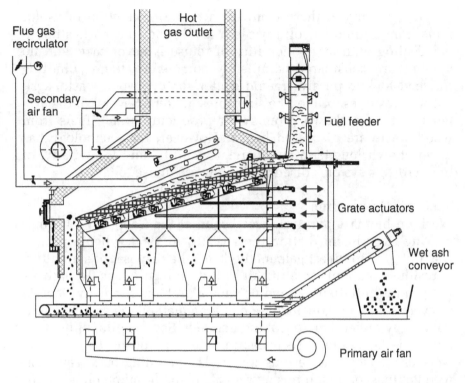

**Figure 4-25.** Reciprocating grate furnace. (Courtesy Enplant Engineering, SA.)

wide size ranges from fines to oversized material, automatic fuel movement and ash removal, and, not least, its robustness. With an adequately sized secondary combustion chamber, combustion can be relatively clean and efficient.

## PYROLYSIS SYSTEMS

Pyrolysis can be defined as burning under less than stoichiometric conditions. The process involves the physical and chemical decomposition of solid organic matter caused by the action of heat in the absence of oxygen. Wood can be pyrolyzed to produce charcoal, and coal can be pyrolyzed to produce coke. Due to the intense heat in a pyrolytic reactor, complex organic compounds can be broken down into simpler chemical products. These products include liquids, gases, and carbon char residue.

Various types of pyrolysis processes have been developed over many years by many different companies. One of the goals sought by many of these companies is the production of useful fuels from municipal solid waste (MSW). The concept has merit, but dealing with large quantities of refuse is never easy, and no systems are considered completely commercial today. Limiting the feedstock to purely organic matter such as wood waste, agricultural residues, or the like simplifies matters. Several of the systems that have shown promise for producing usable fuels from wood waste are discussed below. However, the technology has never shown long-term commercial viability, and few, if any, of the pyrolysis systems discussed below are in use today.

**Tech-Air Corporation**

Work on biomass pyrolysis was begun at the Georgia Tech Engineering Experiment Station (now GTRI) in 1968, when a project was initiated to investigate methods for the disposal of agricultural residues without violating air pollution regulations. In particular, peanut hull disposal had become a problem for many Georgia growers, and it was decided to look for a method to use this potential energy rather than to just incinerate it. Several pilot units were built on the Georgia Tech campus, and pyrolysis experiments were carried out on various materials, including bark, sawdust, wood chips, cotton gin trash, various nutshells, automobile shredder wastes, and municipal wastes. The Tech-Air Corporation was formed to help organize the development efforts. Two field test units (each with a nominal capacity of two tons per hour) were installed at a peanut shelling plant in Georgia. Char and pyrolytic oil were produced successfully with these units, and the experience gained led to the building of a commercial prototype plant in Cordele, Georgia. This prototype plant was operated intermittently for about two years, sometimes on a round-the-clock basis.

During this time, another pilot plant was built on the Georgia Tech campus, designed specifically for the processing of municipal waste. This one-ton-per hour unit was operated on light fraction, heavy fraction (with and without metals), and whole garbage, sewage sludge, and shredded tires. Based on the results of this work, the American Can Company became interested in the process and bought the Tech-Air Corporation in 1975. At that time, two different efforts were initiated to carry forward the work of commercializing the wood-waste system and to establish a continuing research and development program at Georgia Tech in order to support the area of waste utilization.

During the eighteen-month operation at the Cordele plant, the char and oil produced was sold in the bulk char and fuel oil market. The system produced varying amounts of oil, gas, and char according to the control configuration. Energy content of the products, of course, depended on the feedstock, but, in general, the heating value of the char varied from 12,300 to 13,500 Btu per pound. The heating value of the gas produced by the pyrolysis system varied from 3000 to 4000 Btu/lb and the heating value of the oil was generally around 9000 Btu/lb.

The char produced with the Tech-Air system has been used to make charcoal briquettes and has been experimented with in the laboratory to make activated carbon. The pyrolysis oil has been used as a fuel and has shown some potential as a chemical raw material. The oil has been sold commercially for use as a fuel in a cement kiln, a power boiler, and a lime kiln. The pyrolysis oil has also been successfully mixed with #6 fuel oil; however, it had low pH and there were significant concerns about corrosion. Some of the compounds in the oil also polymerized upon heating. Hence, the routine heating and recycling of oil via piping loops between burners and storage tanks for viscosity control, as used for petroleum fuels, is not possible. The gas from the process has been used to fire an infeed dryer at the Cordele plant and, in addition, has been used to fuel an internal combustion engine. The gas had a heating value of up to 180 Btu per cubic foot.

An overall view of the Tech-Air system is included in Figure 4-26. Tech-Air was sold to American Can, and despite the considerable investment and level of technical development, no full-scale plants were sold and the system is no longer on the market.

### Energy Resources Co. Inc.
The Energy Resources Co., Inc. (ERCO) in Cambridge, Massachusetts developed a pyrolysis system using various feedstocks including wood and agricultural wastes. Rather than using a plug-flow fixed bed as in the Tech-Air and Enerco systems, ERCO accomplishes pyrolysis using a fluidized bed. The feed material is introduced into the fluid bed and partially burned to supply the heat for pyrolysis. The remaining solids are partially entrained and carried out of the fluidized bed. These solids contain char particles that are separated from the gas stream with a multiclone.

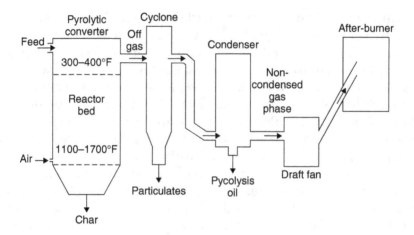

**Figure 4-26.** Tech-Air pyrolysis system.

The char thus collected can be recycled to the fluid bed to produce a higher yield of gas and oil or it can be removed and sold as a feedstock for charcoal briquettes. The oil contained in the gas stream can be recovered by condensation or sent along with the gas to be used for process heating.

The fluid-bed medium is sand and the manufacturer claims that fuel moisture contents of up to 50% can be handled. If this is the case, the ERCO system may have an advantage over the other systems already discussed in that it can burn more relatively wet fuels.

A general view of the ERCO system is shown in Figure 4-27. A complete turnkey operation was built in Belle, Missouri, and there is a plant in Bakersfield, California that fires its boilers on almond shells. This plant had a 20,000 lb/hr steam production rate.

Cost data for wood pyrolysis systems are difficult to obtain as the units under development are still largely in the prototype stages. As can be seen above, the systems tend to be complicated and expensive. The attractive feature of the pyrolysis systems is the possibility for sale of a high-value product (char), but the purchaser of a pyrolysis system should be quite certain about the existence of this market, as existing markets for char and carbon can have well-defined and rigid requirements for feedstock properties. Further information on suppliers of pyrolysis systems is included in Appendix 1.

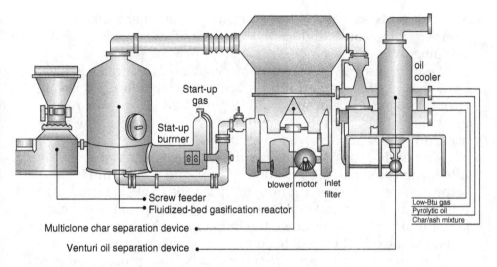

**Figure 4-27.** Fluidized-bed pyrolysis system. [Courtesy Energy Resources Co., Inc. (ERCO)].

## WOOD GASIFIERS

asification can be considered a mature technology as long as the application is one that the vendor has served, successfully, three or more times.

The main problems associated with wood gasifiers involve undesirable constituents in the gas, including tars, acid, and ash. The tars and acids have a tendency to condense in pipes and burners if the gas is allowed to cool or if the gas is not scrubbed as it exits the gasifier. These undesirable properties of the gas and the problems with grate slagging and burnout have been among the most formidable ones facing researchers. Some developers (e.g., EPI, Energy Products of Idaho) have solved the tar fouling problem by simply gasifying it, followed immediately by complete combustion and sending the hot gas to the final application. The only downside of this is the need for higher excess air to keep the ash from slagging.

The basic design for some classes of gas producers dates back to the late 1800s when wood gas was used to fuel engines for power generation and as a substitute for coal gas for lighting and industrial processes. Therefore, due to the age of the basic technology, it has been difficult for researchers and manufacturers to obtain patents as "new" developments. Thus, many small concerns are reluctant to divulge the specifics of systems they are developing.

Gasifier designs include updraft gasifiers, downdraft gasifiers, cross-flow gasifiers, and fluidized-bed gasifiers. Before discussing specific systems in further detail, Figure 4-28 shows some basic principles of gasification. An updraft unit is shown, but the same basic principles are involved in all units.

The updraft gasifier is a simple, older design. The vessel is filled with wood chips, ag residue, or coal, and a small fire is started on a grate above the ash hopper. Just enough air is introduced into the unit to support a glowing bed of coals. Steam may be injected into the bed to promote the formation of hydrogen ($H_2$) and carbon monoxide (CO) and to control grate temperature. In the reduction zone, the hot gases from the combustion process are partially reduced to CO, which is a medium-heating-value gas. The products then pass into a pyrolysis zone where the infeed first

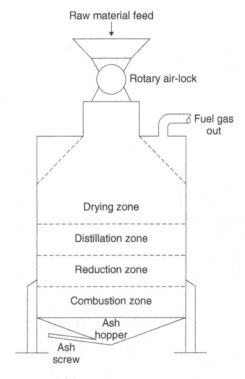

**Figure 4-28.** Updraft wood gasifier.

gives up its volatile gases. At this point, many of the materials that cause problems in the gasifier, such as tars and acids, are produced. In the top section of the updraft unit, a drying process occurs wherein the exiting product gases give up some of their heat to the infeed. The gas is drawn off at the top of the unit and piped to a burner or engine for final combustion. Other methods of gasification differ in regard to where the reactions occur, but the gasification processes are basically all the same. Due to the various gasifier configurations being developed by different manufacturers, the gases produced differ somewhat in their contents. Typical heating value of gas produced in wood gasifiers is 150 Btu/cu ft for a single-chamber unit, with dual-chamber types attaining about double the heating value by reducing $N_2$.

### Biomass Gas & Electric (BG&E)

This firm offers a dual-fluid-bed design gasifier that produces medium-Btu gas. The design is derived from a 1980s Battelle Columbus gasifier design. Past history of the process includes extensive prototype testing at Battelle, and a demonstration installation at a power plant in Vermont.

### Energy Products of Idaho (EPI)

This firm offers single-fluid-bed gasifiers and single-fluid-bed combustors. They have been in the business for more than 20 years. Size ranges are from 30 million Btu/hr up to 500 million Btu/hr (about 400,000 lb/hr of steam equivalent). They have made many systems in the 200,000–300,000 lb/hr range of steam production. Fuels for their systems are wood, biomass, coal, and ag waste.

The EPI approach is a two-stage process in a single chamber: gasifying in the bed, and then injecting air overhead to complete combustion. This eliminates the need for a low-Btu gas burner, but requires close proximity to the boiler or other energy recovery device, and refractory-lined ducts for hot gas transfer. Output to the boiler or dryer is 1400–1800°F to prevent slagging. The low temperature and high excess air reduce overall system efficiency, but lower temperature and lower operating problems increase plant uptime.

A more complete listing of the companies engaged in this field is included in Appendix 1. In general, the main problems with gasification include the following:

1. Lack of long-term operational experience by some manufacturers
2. Potential problems in piping and burners resulting from tars and other liquids in gases
3. Slagging of grates due to ash
4. Requirements for operator attention

One of the most attractive aspects of wood gasifiers is price. As mentioned earlier, packaged wood boilers may cost four to five times as much as natural gas/fuel oil boilers of the same steaming capacity, which have been the mainstay of the commercial market for the past fifty years. Cost figures obtained by the Solar Energy Research Institute and independently by Georgia Tech indicate that gasification systems can be retrofitted to an existing gas/oil burner for substantially less than the cost of a new wood system. Indeed, the major portion of the cost of a gasification system will be for the wood handling, conveying, and metering components. Thus, the industrial plant owner who has a gas/oil boiler with significant useful life remaining will have an attractive alternative to an entirely new system. The cost curves in Chapter 12 illustrate this point more clearly.

## OTHER COMBUSTION SYSTEMS

In addition to the wood burning systems already discussed, there are systems available that do not fit conveniently in the above categories. This hardware is discussed below.

### Heuristic Engineering, Inc.
Heuristic Engineering, Inc. was founded by Dr. Malcolm D. Lefcort, who from 1973 to 1980 was Chief Engineer at Lamb-Cargate Industries Ltd., where the Lamb Wet Cell Burner was developed. Lamb-Cargate went out of business in the mid 1980s, and since then Heuristic Engineering has developed The EnvirOcycler, which is an improved second-generation version of the Wet Cell Burner.

The Lamb-Cargate Company of New Westminster, British Columbia specialized in the supply of hardware for the forest products industry. They developed a system to burn wet wood residue and chips that employed features from several of the categories already discussed. This system was called the Lamb-Cargate Wet Cell and is illustrated in Figure 4-29. The unit functioned partially as a gasifier and partially as a cyclone-type burner. The Wet Cell contained two chambers. Wood residue was fed with an under-feed stoker and was forced up into the center of the horizontal pinhole grate where it formed a conical pile in the primary chamber. Underfire air at 500°F was blown through the bed and the primary chamber functioned as a partial gasifier. The volatiles and fuel gases were driven off and then partially combusted in the top

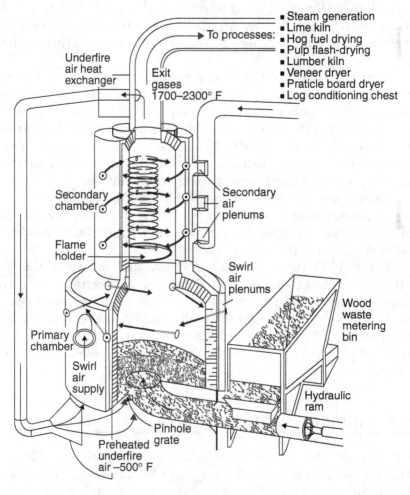

**Figure 4-29.** Lamb-Cargate Wet Cell (U.S. Patent 885-377, Canadian Patent 322-921).

of the primary chamber; hence, this worked like a gasifier and close-coupled combustor. Combustion was completed in the secondary chamber, and gases exited the top of the unit at high temperature. The manufacturer claimed that the unit was capable of burning 65% moisture content (wet basis) wood, and the unit had also been operated on drier wood down to 30%. Depending on the moisture content, the output temperature was reported to be 1700 to 2300°F. The output of the unit could thus be ducted to a waste-heat boiler or various types of dryers. An early production Wet Cell was fitted to a lumber drying kiln in northern British Columbia and reportedly operated satisfactorily. Lamb-Cargate tested a 60 million Btu/hr Wet Cell to be retrofitted to a rotary lime kiln for a paper mill in Vancouver. Auxiliary heat input was necessary to reach the temperature range (2300 to 2700°F) used in this application.

Control of the Wet Cell was accomplished by controlling the amount of air admitted to the lower chamber. This amount was automatically regulated so that the mixture of volatiles and water vapor passing to the second chamber was at an optimum temperature, generally below 1600°F. The secondary control then admitted the necessary combustion air so that the products of combustion left the upper chamber at a steady temperature.

Prototype units were in the 25 million Btu/hr range. Particulate emissions were reported to be "low," but some mechanical collection was required. It appeared that this system would be cost competitive with a wood gasifier for a given output.

The EnvirOcycler, the second generation design of the Wet Cell, also gasifies biomass up to 65% moisture content, on a wet basis. Heuristic Engineering has had units in continuous operation in the field for the last 20 years, and typical combined heat and power (CHP) systems produce between 15 and 150 million Btu/hr of process heat and/or between 5 and 15 MW of electrical power.

## ASH SLAGGING AND SLAG CONROL

Many wood- and ag-fired combustors have experienced slagging of the grates. This is caused by melting of the salts and minerals in the ash. In addition, fouling of heat transfer surfaces by the fly ash may also occur. Both add to operating costs and downtime.

The usual solution to slagging of the grates is to add small amounts of reagents, called "fireside additives" to raise the melting point of the ash. Addressing tube fouling is more complex, and can involve constant addition of reagents, or shocking the system with higher concentrations on a periodic basis to change the nature of the fouling and aid in its removal. Firms such as LBG Industries [3] and Fuel Tech, Inc. have created and sold reagents to fix grate and tube fouling problems for many years.

## Empirical Tests for Slag Control

ASTM methods can be used to determine ash fusion temperature and mineral analysis in slag and fly ash samples. Ash fusion temperatures can be determined under both reducing and oxidizing conditions in the laboratory. Typical critical temperature points used in solids fusion tests are initial deformation temperature (IT), softening temperature (ST), hemispherical temperature (HT), and fluid temperature (FT). In the case in which dry ash is desired, an ash-plus-flux mix that has a high ST through FT temperature would be desirable to prevent slagging. The empirical testing involves adding small amounts of candidate reagents to the ash and firing it in muffle furnaces to optimize the reagent and the concentration.

## Research on Controlled Melting

Pilot-scale research been carried out to change the fluxing properties of solids processed in high-temperature rotary kilns. This work was done by U.S. EPA's SITE (Superfund Innovative Technology Evaluation) Emerging Technology Program [4]. For this project, the goal was to trap toxic metals (such as cadmium, chromium, and lead) that had been spiked into a synthetic soil matrix feedstock made of sand and kaolin clay. Although the normal kiln discharge is a dry dust, use of small amounts of reagents resulted in strong, durable nodules with excellent crush strength and improved resistance to leaching. Feed preparation, particularly control of moisture content, was found to be a key element in initiation of agglomeration of the feedstock upstream of the kiln. A good correlation was found between decreasing metals leachate levels and increasing crush strength. Table 4-2 shows the fluxing reagents used.

**Table 4-2.** Fluxing compounds

| Flux | Compound |
|------|----------|
| Iron | $Fe_2O_3$ |
| Calcium | $CaO$, $Ca(OH)_2$ |
| Silica | $SiO_2$ |
| Magnesium | $MgO$ |
| Sodium | $Na_2CO_3$ |
| Potassium | $K_2CO_3$ |
| Alumina | $Al_2O_3$ |
| Titanium | $TiO_2$ |

### Slurry Treatment in Power Boilers

Fuel Tech supplies equipment and reagents to modify fouling on boiler tubes of power boilers. A MgOH slurry is injected into the furnace. The particles produced by vaporizing the slurry in the combustion zone have very small particle size and a very high surface/volume ratio. Dosage rates are in the range of 1–2 lb per ton of coal. The MgO does not affect slagging temperature, but rather weakens the fouling and promotes removal via soot blowing. Fuel Tech promotes use of its "Targeted In-Furnace Injection" system, which uses CFD to model the furnace and optimize injection points [5]. Use of MgO, or $Ca(OH)_2$ for slag modification also has the benefit of reducing $SO_2$ and $SO_3$ by producing Mg or Ca sulfates as a reaction product.

Fuel Tech's TPP-555 [5] notes that traditional injection of reagent with fuel requires higher rates of addition, and that traditional reagent add rates were 3:1 Mg to V (vanadium)for No. 6 fuel oil.

Other reagents have been used. Fuel Tech suggests use of vermiculite for slag and fouling modification, and others note use of $Ca(OH)_2$ injection into the combustion chamber, as well as use of high-conductivity refractory in the combustion chamber [6].

### Tube Fouling and Subsequent Corrosion in MSW Boilers

Whereas fouling reduces heat transfer and can cause boiler tube failure due to stress and hot/cold spots on the tubes, fouling-induced corrosion is also an issue.

MSW (municipal solid waste) boilers have problems with tube fouling due to soft, semimolten slag particles adhering to boiler tubes, and gaseous metals and salts condensing on much colder boiler tubes. The ash is a combination of that found in cellulosic feedstocks combined with a wide range of ash constituents from other waste. Research has been done on these slags and, as noted in [6], metals in MSW boilers corrode at high temperatures due to attack by fouling. The principal problems are oxidation by metal oxides, sulfidation by metal sulfides, sulfidation/oxidation by mixtures of sulfides and oxides, carburization by metal carbides, and chlorination of metals by metal chlorides.

Corrosion due to fouling may occur with wood and ag feedstocks, but the chemistry is different from that of MSW slag. Wood and ag fuels generally are low in chlorine; however, some contain chlorine from sea salt due to dragout of wood near the ocean. Sulfur is usually low also, but some sulfates will be present, and Na and K will combine with sulfur to form sulfates.

## REFERENCES

[1]  Levi, M., P., et al., *Decision Makers Guide to Wood Fuel for Small Industrial Energy Users,* National Technical Information Service, Springfield, VA, 1980.

[2]  *Power Engineering,* June 2007.

[3]  www.lbgindustries.ca.

[4]  McGowan, T., R. A. Carnes, J. N. Lees, G. A. Heian, and M. K. Richards, "Thermal Encapsulation of Heavy Metals," presented at AWMA conference, June 19–24, 1994.

[5]  Bulletin TPP-555 and 563, Fuel Tech, Inc. , www.fueltechnv.com.

[6]  High-Temperature Corrosion in Waste-to-Energy Plants," Lee, S.-H., Themelis, N. J., and Castaldi, M. J., *Journal of Thermal Spray Technology,* Vol. 16, No. 1, March 2007.

# CHAPTER 5

# BIOMASS FUEL STORAGE AND HANDLING

## SOLIDS HANDLING FUEL PROPERTIES

The editor of this book started his career in solids handling and can attest to the fact that gases and liquids are easier to handle than solids. For solids, there is no substitute for experience and using vendors that have handled the material before, or something very close to it in density, angle of repose, size, and other important properties.

For material handling purposes, biomass, whole green tree chips, bark, and sawmill residues act (to a large degree) more like soil than a free-flowing substance. For solid biomass fuels, primary characteristics of concern during storage and handling are its moisture content and size.

Wood is the most widely used biomass fuel. It can be a waste product of lumber or furniture manufacturing. It can also come from waste generated by conventional harvesting techniques or it can be directly harvested from the forest. Most available wood fuels are green. Dry wood fuel comes from manufacturing (such as furniture operations, sander dust, and planer shavings) that uses dried lumber wood. The choice between dry or green wood fuel for a particular location is decided on the basis of availability and cost. One consideration is that combustion equipment for green fuel can also burn dry fuel, but the reverse is seldom true.

The size of wood fuel is also important and affects handling, storage, on-site fuel preparation, combustion systems, and emissions. There are no official standards for wood fuel; industry practice is to analyze availability of wood fuel and design the system accordingly. The 2″ × 2″ × ¼″ chip is frequently used for systems fueled from whole tree chippers. Other ranges are sander dust up to ¼" for suspension burning, hogged fuel with a range of sizes and top size of 2", and mill yard waste from sawdust to cants and boards several feet long. The latter would have to be chipped, hammermilled, or hammer-hogged to smaller sizes either at the producer or at the user end.

Another general classification of wood fuel that is available in certain areas is "bark and shavings," derived from debarking of logs. Depending upon the maximum size of the bark and shavings, this fuel may require size reduction. Bark fuels tend to be high in silica (sand) content. Sand enters the bark in two ways: (1) in coastal regions, sand is embedded into the surface layers through wind transport; (2) during skidding (especially in rainy seasons), sand (and clay) is picked up by the surface layers. The high silica content (noncombustibles may reach 14%) creates problems in clogging of boiler grates, erosion of fuel and ash handling systems, heat exchanger wear, and increased particulate emission.

Wood fuel for industry can be purchased on several bases, but the common method is by the ton. Some contracts have an adjustment for moisture content as surrogate for heating value. This is not usually critical in that the various green wood materials tend to be plus or minus 5% of 4000 Btu per pound. For dry wood, more care must be exercised because the percentage of drying affects the heat content. For example, kiln-dried material may have a 10% moisture level and contain 7200 Btu per pound. Dry, light fuels such as planer shavings are rarely shipped far due to low density and high transport costs.

Wood pellets are occasionally used as industrial fuel, and pellets made in Georgia are finding their way to Scandinavia to be burned in utility boilers. The cost is very high, and use of pellets is due to compliance with Kyoto greenhouse gas protocols rather than cost savings. Pellets are made by drying the wood, reducing it to a small particle size, and then densifying it by extrusion through dies. This process produces a product in the form of pellets or cubes. The bulk density is about 35 lb/ft³ compared to about 23 lb/ft³ for wood chips. The uniform product allows for easy storage, handling, and use in combustion equipment; however, dusting can be a problem. Combustion thermal efficiency can be 15% higher than obtainable with green wood, and the capital cost of the combustion equipment can be reduced. The costs of

densification are significant, however, so the place of manufactured wood fuel is not clear at this time. Typical selling prices are
in the range of $150/ton, several times the cost (on a Btu basis) of
wood residue.

In order to plan the size of wood yard required, the boiler fuel
requirements must be determined. Using Table 5-1, the calculations for determining the quantity of fuel required are:

Quantity of fuel required = steam demand × enthalpy of steam
× 1/100 V of wood × 1/boiler efficiency

where:

Steam demand is in lb-steam/hr
Enthalpy of steam (at 150 psig from 220°F feedwater) = 1000 Btu/
lb-steam
HHV of the specific wood fuel, as specified
Boiler efficiency = 65% (typical for green wood fuels)

For a 1000 hp boiler burning whole-tree chips (50% moisture
content wet) at 65% efficiency, the hourly fuel requirement is:

1000 boiler hp/hour × 34.5 lb-steam/boiler hp
× 1000 Btu/lb-steam × 1 lb of wood/4000 Btu/lb
× 1 ton/2000 lb × (1/0.65 boiler efficiency
= 6.6 tons/hour of green whole-tree chips

For some commonly used industrial boilers, Table 5-2 shows calculations of quantity requirements using green wood chips as fuel
and a boiler combustion efficiency of 65%. This table assists in

**Table 5-1.** Wood fuel properties

| Wood fuel | Moisture content (wet basis) | Heating value (Btu/lb) | Bulk density (lbs/ft³) |
|---|---|---|---|
| Whole-tree chips | 50% | 4000 | 24.0 |
| Green sawdust | 50% | 4000 | 20.0 |
| Dry planer shavings | 13% | 6960 | 6.0 |
| Dry sawdust | 13% | 6960 | 11.5 |
| Wood pellets | 10% | 7200 | 35.0 |

**Table 5-2.** Wood fuel quantity requirements

| Boiler hp | Steam output (lb/hr) | Wood consumption | | | Truck loads/day (23 tons/load) |
|---|---|---|---|---|---|
| | | Tons/hr | Tons/24 hrs | ft³/hr | |
| 100 | 3,450 | 0.67 | 16.08 | 55.83 | 0.75 |
| 200 | 6,900 | 1.35 | 32.16 | 111.67 | 1.5 |
| 300 | 10,350 | 2.01 | 48.24 | 167.49 | 2 |
| 400 | 13,800 | 2.70 | 64.32 | 223.33 | 3 |
| 500 | 17,250 | 3.35 | 80.40 | 279.17 | 3.5 |
| 600 | 22,500 | 4.02 | 96.48 | 334.98 | 4 |
| 700 | 24,150 | 4.69 | 112.56 | 390.81 | 5 |
| 800 | 27,600 | 5.40 | 128.64 | 446.67 | 6 |
| 900 | 31,050 | 6.03 | 144.72 | 502.47 | 7 |
| 1000 | 34,500 | 6.60 | 160.80 | 558.33 | 7 |

Notes: 1. Fuel is green wood chips. 2. Boiler combustion efficiency assumed to be 65%.

calculations necessary for sizing wood fuel handling and storage facilities.

## RECEIVING

There is no one optimum method by which biomass can be received and unloaded at an industrial plant. It is necessary to evaluate specific conditions at each site. The capabilities of prospective wood fuel suppliers also influence the decision. Economics is a major consideration, and the lowest cost method that meets the particular needs should be selected. Figure 5-1 illustrates four receiving methods that are widely used, and Table 5-3 shows estimated costs per year.

### Live-Bottom Van

The live-bottom van is a self-unloading trailer with a conveyor incorporated into the floor. A typical van is 40 ft long, 8 ft wide, and 13 ft 6 in high, and will transport about 23 tons of material per load. The conveyor travels at the rate of six feet per minute, permitting the van to be unloaded in less than 10 minutes. Live-bottom vans cost approximately $60,000, which is twice the cost of standard open-top trailers of the same size. The costs incurred to operate the van are reflected in increased transportation charges

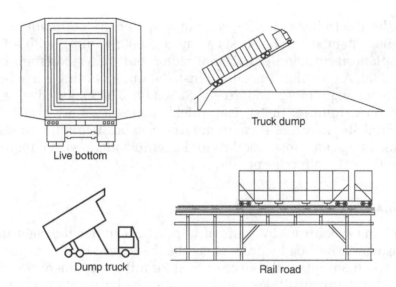

**Figure 5-1.** Wood fuel receiving methods.

on the order of 10% to 15%. The advantage to fuel purchasers of live-bottom van deliveries is that no costly unloading device is required on-site. This may be an appropriate method for six deliveries per day or less.

## Truck Dumps

Truck dumps are devices installed at the plant site that elevate and unload trailers. Most are hydraulically operated. They come in two sizes: one handles both tractor and trailer and the other

**Table 5-3.** Comparisons between unloading alternatives (estimated cost per year)

| Boiler size | | | Live-bottom | | Front-end |
|---|---|---|---|---|---|
| hp | lb/hr | Tons/yr | van | Truck dump | loader |
| 100 | 3,450 | 4,900 | $2,100 | $27,000 | $10,000 |
| 1000 | 34,500 | 49,000 | $21,000 | $29,000 | $43,000 |
| 3000 | 100,000 | 142,000 | $61,000 | $34,000 | $108,000 |
| 7500 | 250,000 | 355,000 | $153,000 | $70,000 | — |

Note: 1984 costs escalated for CPI inflation via 215.6/103.9 = 2.1 factor.

handles the trailer only. A truck dump with a 40-foot platform to handle a standard trailer costs about $75,000; a dump with a 60-foot platform to handle both the tractor and trailer costs around $175,000. A complete (installed) installation, including a live-bottom receiving hopper, can exceed $250,000. The major advantage of a truck dump is that it can quickly unload widely available, standard 40-foot vans. Turnaround time can be 20 minutes or less. Truck dumps are most practical for larger installations that require more than 10 deliveries per day.

### Railroad Delivery

Receiving wood fuel by railroad is practical under the right circumstances. Railroad shipment can be the lowest cost transportation when shipping distances exceed 70 miles and quantities are large. If a railroad siding is available and the local area wood fuel is in short supply, then railroad shipment should be evaluated.

### Dump Trucks

Long-body dump trucks are sometimes used for wood fuel delivery. Though usually of smaller capacity than standard 40-foot vans, they can be utilized in specific circumstances for short-haul situations and low wood use rates. Standard dump trucks are used, but are limited to short-haul situations.

Small installations not having specialized unloading devices could employ a small front-end loader for this purpose. Front-end loaders can unload trailers in around 30 to 45 minutes. When the tractor operation and labor costs are considered, this method can be justified economically only in a few small-size operations. Unless the operator is skilled, damage to the trailer can occur.

## STORAGE

Storage of ag residues, if wet, can be problematic due to biological action. For example, silage gives off nitrogen oxides which have proved lethal to those entering confined spaces without sufficient ventilation. Dry ag residue is more stable and less problematic, but requires covered storage or silos to prevent moisture gain due to rain.

Storage of wood fuel is one of the first considerations in the design of a wood fuel facility. Often, a significant land area is required, and the location of available storage area relative to the combustion burner location affects the fuel handling system design and costs.

The size and type of wood storage required varies with the requirements of a particular plant. Among the factors to be considered are:

- Heat energy requirements of the plant
- Land area availability and its location
- Fuel moisture content
- Fuel preparation requirements
- Reliability of wood fuel supply system
- Weather severity

All factors should be evaluated and considered in the storage system. The actual storage is usually accomplished by open storage, covered storage, or by silos. Figure 5-2 provides schematic views of typical storage methods.

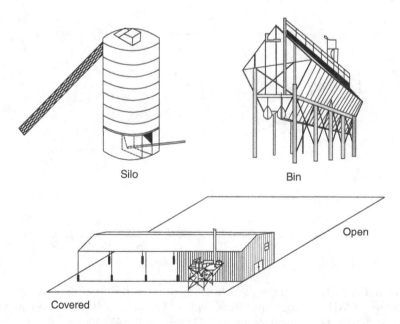

Silo

Bin

Open

Covered

**Figure 5-2.** Wood fuel storage methods.

## Open Storage

Open storage is an area that is not covered for protection from rain or other precipitation. It can be the only storage needed, but is most often used for a 10 to 30 day supply to handle interruptions in the normal fuel delivery system such as inclement weather or strikes. The quantity of material storable on a given site is limited to the height obtainable by the stacking and retrieval equipment.

It is best to pave open storage, preferably with concrete (using coated rebar for reinforcement), to prevent the accidental scraping up of dirt with the fuel, which can cause problems with the handling and combustion equipment. The area should be pitched to allow for drainage and should not be located in a flood-prone area.

All wood fuels undergo losses in net available energy as a function of storage time (see Figure 5-3). "Lifo" (last-in, first-out) rather than "fifo" (first-in, first-out) fuel utilization procedures are recommended.

The primary cause of depletion of available energy (Btu/lb as fired) of openly stored wood fuel is an increase in the surface layer moisture content due to precipitation. Once saturation of the surface volume is achieved, future weather conditions will not affect

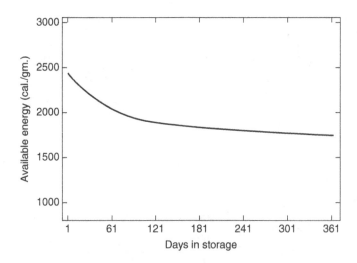

**Figure 5-3.** Available energy as a function of storage time for wood chips. (Courtesy of M. L. Curtis; from "The Effects of Outside Storage on the Fuel Potential of Green Hardwood Residues," 1980, M.S. Thesis, Virginia Polytechnic Institute and State University, Blacksburg Virginia 24061.)

moisture content in this outer zone. Due to the high packing density and minimal air flows through most piles, the wood within about 1½ to 2 ft of the surface will remain saturated. Piles constructed in shapes having large surface volume/total volume ratios (i.e., rectangular piles) usually display net increases in average moisture content through the pile. Construction of conical piles will minimize surface volume/total volume ratios and consequent increases in net moisture content, and will thus minimize any decreases in available energy.

The secondary cause of available energy depletion is loss of volatiles and biochemical oxidation, accounting for perhaps 15% of loss in available energy. Such losses are tempered by the poor air flow characteristics typical of most piles (Figure 5-4). This "chimney effect" contributes to high internal pile temperatures (Figure 5-5). Of course, the high internal pile temperatures can contribute to significant moisture content decreases in the central zones (Figure 5-6).

Strongly influencing the natural drying potential of stored wood fuel is the packing density of the pile. The packing density is the configuration of the discrete wood chips and/or grains in the as-constructed state. Piles constructed with bulldozers or like equipment will have higher packing density and inferior air flow characteristics relative to those constructed by the gravity method. It can be expected that gravity constructed piles will maximize the

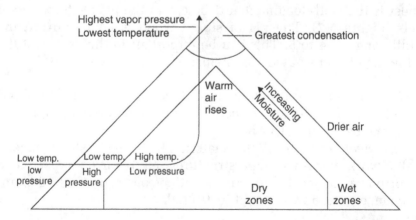

**Figure 5-4.** Chimney effect in storage piles. (Courtesy of M.L. Curtis; from "The Effects of Outside Storage on the Fuel Potential of Green Residues," 1980, M.S. Thesis, Virginia Polytechnic Institute and State University, Blacksburg, VA.)

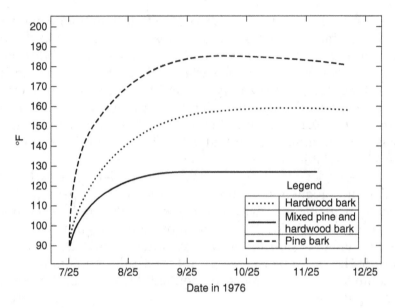

**Figure 5-5.** The maximum internal temperatures measured within experimental storage piles of residues as a function of storage time. (Courtesy of M.S. White, from "Bulk Storage Effects on the Fuel Potential of Sawmill Residues," 1978, *Forest Products Journal,* Vol. 28, No. 11, pp. 24–29.)

drying effects of naturally occurring high internal pile temperatures.

Another aspect of long-term storage important to the plant engineer is the well-documented decrease in the pH of stored wood fuels (Figure 5-7). The pH of stored wood is commonly in the acidic range of 4 to 5. This can be important in the design of the following portions of wood systems:

1. *Concrete slabs or structures* in intimate contact with the wood fuel (use of acid-resistant concrete mix and epoxy-coated reinforcing bars is effective here).
2. *Any steel portions* of the material handling system (augers, ductwork, conveyor superstructures, etc.) can be expected to corrode at an accelerated rate if anticorrosion/preventive maintenance procedures are not instituted.

Long-term open storage of green whole-tree chips, residues, and bark necessarily entails comprehensive safety procedures. The highest internal pile temperatures are found in pine bark

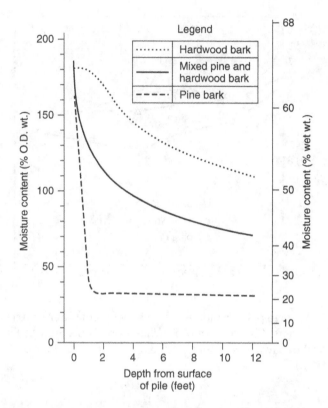

**Figure 5-6.** Moisture content as a function of depth of green bark and sawdust piles. (Courtesy of M.S. White, from: "Bulk Storage effects on the Fuel Potential of Sawmill Residues," 1978, *Forest Products Journal,* Vol. 28, No. 11, pp. 24–29.)

piles, followed by hardwood bark and mixed pine and hardwood bark (Figure 5-5).

With due consideration of safety, space, and cost efficiencies, the following storage procedures are recommended.

### For Sawdust and Mill Residues

When natural drying is not a criterion, it is safe and practical to construct large (in area extent), high piles compacted with heavy equipment such as loaders, bulldozers, etc. Installation of one Type T thermocouple per six feet of pile height in cold climates (see discussion below) will minimize fire hazards. Should pile temperature exceed 170°F, danger of fire is imminent and the pile should be dismantled.

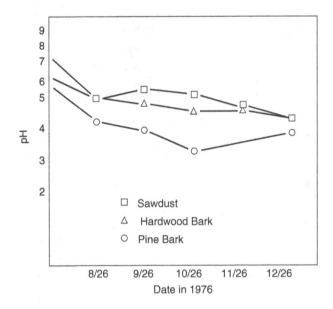

**Figure 5-7.** Average pH of residues within experimental storage piles as a function of storage time. (Courtesy of M.S. White from: "Bulk Storage Effects on the Fuel Potential of Sawmill Residues," 1978, *Forest Products Journal,* Vol. 28, No. 11, pp. 24–29.)

When natural drying of stored fuel is the desired goal, build gravity-constructed nonpacked piles. Thermocouple installation is recommended; however, elevated temperatures should not be expected.

**For Green Whole-Tree Chips and/or Bark**
As the percentage of bark in a pile increases, so will internal temperatures. The occurrence of high pile temperatures in bark (and to a lesser extent whole-tree chip) piles is well documented (Figure 5). Therefore, long-term storage piles of bark and/or green whole-tree chips should be gravity constructed, *not* compacted with heavy machinery, and instrumented with thermocouples, one for every six feet of pile height. Again, should pile temperatures exceed 170°F, danger of fire is imminent and the pile should be dismantled.

The greatest danger of fire exists under conditions that restrict and/or eliminate air flow through the pile (and consequent internal-pile heat dissipation). A well-documented cause of spontaneous fires occurs during the winter months when a layer of ice can form over the outer pile surface. During such periods, internal pile heat cannot be dissipated and fires often result. High concentrations of bark fines at the pile surface can also replicate this condition. Close observation of internal pile temperatures during these periods will help ascertain imminent danger.

For an exhaustive state-of-the-art discussion on the above subject, the reader is referred to "The Effects of Outside Storage on the Fuel Potential of Green Hardwood Residues," by Michael L. Curtis [1]. The industry has standards on limiting pile height and maintaining proper distances between outdoor storage piles to allow for both firefighting and natural ventilation and heat dissipation.

As stated, the storage volume per unit area is determined by pile height, and the total weight is obtainable by multiplying the volume by the specific weight of the particular wood fuel. As an example, if whole-tree chips at 23 lbs/ft$^3$ are stacked to an average height of 12 ft, then 138 tons can be stored per thousand square feet of area. The cost of paved open storage depends upon local conditions (amount of area preparation necessary), but an average cost of $3.60 per sq ft (2008 dollars) can be used for preliminary estimating purposes.

## Covered Storage

Wood storage systems frequently contain a covered storage area, using an open-sided metal building located adjacent to the combustion equipment. It is sized for a three- or four-day storage capacity. Many of the same considerations considered under the discussion of open storage apply to covered storage.

Fuel to be immediately burned is taken from this area. The cover prevents soaking by rain and provides better conditions for operation of the front end loader. Under normal operation, most of the fuel can be delivered to the covered storage area and consumed without going through the open storage area.

Covered storage is necessary for dry fuel, such as many ag fuels, fuel from kiln dried lumber, or prepared, densified wood. For estimating, Figure 5-8A shows storage capacities for whole-tree chips and the typical costs of covered storage sheds.

## Silos

Silos are used to store wood fuel under certain conditions. For the nonforest industry, it is expected that silos will be considered when:

1. Fuel requirements are relatively small, about one fuel delivery per day or less
2. Automated fuel feed is highly desirable
3. Site location prevents the use of open storage

Silos come in a variety of diameters, heights, and volumes, and they can be manufactured of metal, poured concrete, or staved concrete.

Wood fuel as a class has poor flow characteristics, and this must be considered in the silo design. Wood fuel may not be free flowing in silo storage; therefore, silos require active mechanical retrieval systems. These are manufactured in several designs and can include a chain flail or screw auger. Wood fuel in silo storage can bridge over (Figure 5-8B). Under certain conditions, the collapse of this bridge can create outward thrust and/or vacuum conditions severe enough to cause structural failure of the silo.

There are a variety of successful wood fuel silo storage systems now in operation with varied shapes (round silos and rectangular silos with negative wall angles, growing larger from top to bottom) and types of unloaders. Properly designed, they can achieve full automation.

Table 5-4 presents some typical steel silos showing size, storage capacity, and approximate installed costs. Concrete silos are about 12% less on a first cost basis.

## IN-PLANT FUEL HANDLING

This section describes the various methods commonly employed in the transport of wood fuel at a plant site. After delivery, the wood may be transported to long- or short-term storage, to fuel preparation equipment, or to a hopper at the combustion device. A key design element of wood handling systems is minimization of the amount of transport equipment.

Viable transport methods include:

• Belt conveyors
• Chain conveyors
• Augers
• Front-end loaders

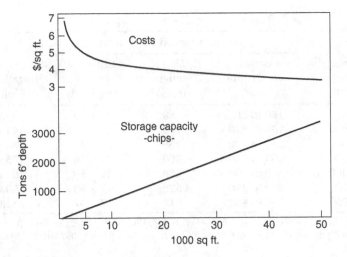

**Figure 5-8A.** Covered storage costs and capacities. (Courtesy of Curtis, G.B., Brown, M.L., and others, "Wood Energy Economics—Proceedings, April 30, 1980," Georgia Institute of Technology, Engineering Experiment Station, April 1980.)

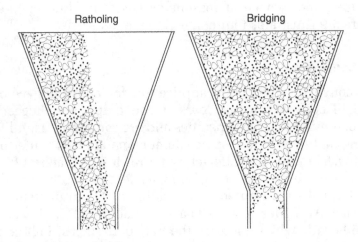

**Figure 5-8B.** Wood flow handling problems. (Courtesy of Committee on Industrial Ventilation, Industrial Ventilation—A Manual of Recommended Practice, American Conference of Government Hygienists, Lansing, MI, 1978.)

**Table 5-4.** Steel silos with unloader [2]

| Storage Capacity (ft³) | Weight of wood fuel stored | | | |
| | Green wood chips (tons) | Wood pellets (tons) | Silo size (dia. × height: ft) | Installed cost (dollars) |
|---|---|---|---|---|
| 5,000 | 50 to 60 | 88 | 15 × 38 | 95,000 |
| 10,000 | 100 to 120 | 175 | 21 × 33 | 111,000 |
| 15,000 | 150 to 180 | 263 | 21 × 48 | 130,000 |
| 20,000 | 200 to 240 | 350 | 21 × 60 | 155,000 |
| 25,000 | 250 to 300 | 438 | 27 × 50 | 165,000 |
| 30,000 | 300 to 360 | 526 | 27 × 58 | 190,000 |
| 35,000 | 350 to 420 | 613 | 27 × 67 | 212,000 |

Notes: 1. Silos are constructed of steel with fused glass. 2. Height is from floor to top. 3. Bulk density of wood is assumed to be 23 lbs/ft³. 4. Bulk density of pellets is assumed to be 35 lbs/ft³.

In addition, there is limited application of pneumatic conveying, vibrating conveyors, and bucket elevators.

The demarcation between various transport methods is not always clear, and their recommended uses can overlap. Careful attention is necessary at final design to select the system that will provide good service at reasonable cost with low maintenance. Figure 5-9 illustrates four conveying methods.

### Belt Conveyors

Belt conveyors find wide application for conveying wood fuel. Used to convey sawdust, bark, wood chips, and hogged wood waste, they have large capacities and are especially useful for conveying fuel over long distances. Belts operate best under uniform loading. They are not satisfactory for retrieving material from storage where the material may be many feet in depth over the belt. Usually, belts are limited to an incline of 15° to prevent material slippage. At higher cost, belts are available with flights and flexible sidewalls which increase the transport angles. Belt conveyors come in various sizes, widths, and lengths, from a few feet up to thousands of feet.

### Drag Chains

Drag chain conveyors are widely used where the transport distance is relatively short. Though construction details differ, they

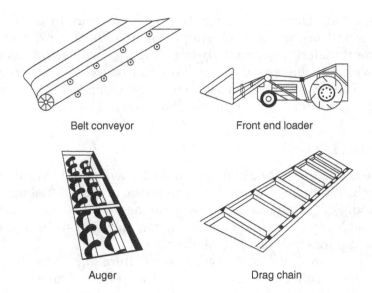

Belt conveyor          Front end loader

Auger                  Drag chain

**Figure 5-9.** Wood yard fuel handling systems. (Courtesy of Clifton, David S., Jr., and Bulpitt, William S., "A Feasibility Study for Wood Energy Utilization in Georgia, Final Report, Project A-2140," prepared for The Georgia Forestry Commission under the sponsorship at The Coastal Plains Regional Commission, Georgia Institute of Technology, August 1979.)

essentially consist of one or more metal chains connected with drag bars running in troughs. Used to transport sawdust, bark, wood chips, and hogged or unhogged wood pieces of moderate size, they can be installed in parallel to create live-bottom retrival systems. The are low-speed, high-load devices and should be provided with shear pin and/or motor overload safety devices to address jamming conditions. A contact in the industry recounted the tale of a lattice-boom-type frame drag chain that proceeded to shorten itself when it jammed due lack of overload protection.

Drag chains are rugged in construction and can handle heavy nonuniform loads. They are especially suited for reclaiming wood from receiving hoppers or storage piles, and for feeding combustor hoppers. They are relatively easy to operate and maintain. Broken links can be replaced or welded by plant personnel.

**Augers**

Augers or screw conveyors consist of a helical blade mounted on a shaft enclosed by troughs or covers that hold the material to be transported. They are commonly used in feeders to the burners and as part of the retrieval equipment in silos or bins. They can transport material horizontally or on inclines with reduced capacity. With variable speed drives, they can be used as metering devices. They can handle hogged wood and wood chips. Augers are inexpensive and easy to operate if run less than 40% full.

### Front-End Loaders

Front-end loaders are useful machines for wood fuel handling. For many installations, the machine can be the principal fuel handling method. For installations that use systems such as drag chains and belt conveyors, the front-end loader can clean up spills and serve as backup for equipment breakdown.

There are two basic types: agricultural and construction. These units differ primarily in their design duty. The agricultural tractor is lightweight and is not expected to maintain a heavy-duty cycle, whereas construction units often are of articulated design for maneuverability and continuous duty. For wood fuel use, the standard buckets should be replaced with larger volume buckets. Wood fuel at less than 25 lb/ft$^3$ can be easily handled using high-volume buckets, known in the trade as snow buckets or loose material buckets.

As a size selection aide, Table 5-5 has been prepared. The table is based on the loader being able to move twice the required weight of fuel per hour, that is, the completion of two cycles of

**Table 5-5.** Front-end loaders for wood-fired boilers [2]

| Type | Bucket size (yds$^3$) | Applications boiler size up to | Tractor wt (lb) | Estimated fuel consumption (gal/hr) | Engine (hp) | Approx. cost (dollars) |
|---|---|---|---|---|---|---|
| Agricultural | 1½ | 600 hp or 21,700 lb/hr | 6,300 | 1.0 | 45 | 40,000 |
| Agricultural | 3 | 1000 lb or 34,500 lb/hr | 7,800 | 1.5 | 64 | 55,000 |
| Construction | 5 | 3333 hp or 115,000 lb/hr | 21,000 | 3.5 | 80 | 140,000 |
| Construction | 12 | 6666 hp or 230,000 lb/hr | 37,000 | 7.5 | 170 | 320,000 |

scoop, move, dump, and return. Taking into account their more substantial design, a 75% duty cycle is applied to the construction loader, whereas a 25% duty cycle is applied to the agricultural loader.

## Other Methods

Among other wood fuel transport methods are pneumatic conveyors, vibrating conveyors, and bucket elevators. Pneumatic conveyors are used in the lumber and furniture manufacturing industries to pick up sawdust for transport to central storage. As operating costs are relatively higher than belt conveying, limited use is expected in the nonforest industry. Vibrating conveyors have certain special applications, but are not widely used. Bucket elevators are modified chain conveyors with buckets attached and are useful where vertical lift is required.

## PREPARATION OF WOOD FUEL

The preparation of wood fuel at a nonforest industry plant may be necessary to facilitate handling and to meet the requirements of the burning system. It is normally limited to size reduction and the removal of foreign objects. In some instances, predrying may be appropriate where waste heat is available or where higher flame temperatures are required for processes such as calcining. Many nonforest industries purchase their fuel in a form such that no on-site preparation is required.

## Size Reduction

Size reduction is required when the wood fuel has dimensions too large to pass through the burner feed system or to combust properly. For example, two inches may be the maximum dimension desired. If any of the wood fuel exceeds this dimension, it can be reduced to proper sizes by passing through a machine commonly called a hog, knife hog, or hammermill.

To minimize first cost and operating expense, on-site preparation of wood fuel should be kept to the minimum consistent with the needs of the combustion system. Therefore, prior to the purchase of a particular unit, the availability of suitable wood fuel should be evaluated. For example, some wood-burning units require dry or partially dried wood. If only green fuel wood is available locally at reasonable cost, then only systems that are capable of using green wood should be considered.

Figure 5-10 [2] illustrates the arrangement of a typical size-reduction system for waste wood. The unhogged material first passes under a magnetic metal separator that removes tramp iron or steel. There are several designs of magnetic separation devices, but all work on the principal of the magnetic attraction of iron. Metal detectors can be mounted under belts to stop the belt when metal is detected that gets past the magnet. After the metal detector, the material then passes onto the top of the disk screen. The disk screen consists of a series of disks rotating on parallel shafts. The spacing of the disks is such that material over a certain size (e.g., 2″) will pass across the top of the disk screen and drop into the hog. The hog will then reduce the larger pieces to the required size (e.g., 2″). The purpose of the screen is to reduce the horsepower requirements of the hog by screening out the material that is of proper size. In the usual design, the material passing through the screen and the hog is forwarded onto a conveyor and transported to storage. Costs of some typical screens, hogs, and hammermills are listed in Table 5-6. Vibrating screens, also called shaker screens, are sometimes used in place of disk screens.

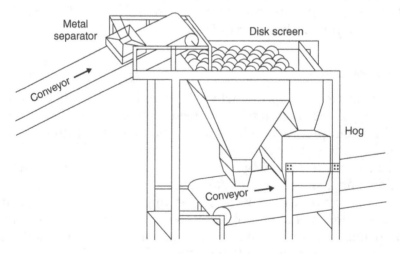

**Figure 5-10.** Wood fuel size reduction system.

**Table 5-6.** Wood fuel size reduction equipment: costs

| | Capacity (tons/hr) | Cost (dollars) |
|---|---|---|
| Screen Costs [2] | | |
| Small disc | 40 | 50,000 |
| Large disc | 250 | 130,000 |
| Small shaker | 10 | 14,000 |
| Large shaker | 40 | 51,000 |
| Revolving drum | 40 | 110,000 |
| Costs of Hogs and Hammermills | | |
| Small-knife hog | 10 | 30,000 |
| Large-knife hog | 32 | 95,000 |
| Small hammermill | 3 | 25,000 |
| Large hammermill | 8 | 61,000 |

## DRYERS

Drying wet wood or ag feedstocks allows higher flame temperatures of 2300 to 2500°F to be attainable (instead of 1800°F possible
with green wood). Also, the overall thermal efficiency of a boiler
can be increased (5 to 15%). Unless there is "free" heat to do the
drying, however, there is no overall energy benefit, as it can be ar-
gued that green wood is "dried" in the furnace as part of the combustion process.

### Pile Drying

Research done by White and DeLuca of Virginia Polytechnic Institute indicates that the elevated temperatures associated with bulk
storage woodpiles result in significant moisture content reduction,
coupled with a probable net decrease in pile heat content. Piles of
different geometries were investigated. The piles were variously
mixed green sawdust, hardwood bark, and pine bark. All piles ex-
hibited a rapid increase in internal pile temperature, with the
greatest temperature near the geometric center at the base of the
pile. The temperatures rapidly reached a threshold value and
thereafter remained fairly steady, varying from steady-state levels
of 120 to 170°F. Although some decrease in moisture content was
observed, this research is not conclusive. Gains in net heating content due to moisture content decrease during open pile storage periods are tempered, and may well be offset, by saturation of the

surface layer by precipitation and evaporation of volatiles. Piles with large surface volume/total volume ratios (i.e., "small" piles) are particularly sensitive to surface saturation from precipitation (see Figures 3-3, 3-5, and 3-6). Finally, there is no free lunch, as the heat generated by the pile, which causes the drying, had to come from the pile; hence, it is no longer available for combustion.

## Drying Equipment

A wide range of gaseous-, liquid-, and solid-fuel dryers have long been used in many industries. Single-pass and triple-pass fossil fuel fired, motor driven, rotary dryers are commonly used in fertilizer plants and grain and other process drying applications. Cascade dryers and flash-type dryers are in widespread use in many processes. With the spiraling cost of energy, equipment vendors and biomass users have adapted these existing technologies to drying wood and ag feedstocks. In applications in which exact moisture contents are not required by the process and the existing boiler has solid-fuel firing capability and is presently firing fossil fuel, waste heat derived from flue gas is a demonstrated economical drying agent with short paybacks and good returns on investment. Paybacks on retrofitted waste-heat dryers for systems already burning high-moisture-content wood will tend to be not so dramatic as those calculated for displacement of fossil fuel by dry wood fuel. Keep in mind that boiler flue gas is low in temperature compared to the normal hot gas temperature used in direct firing. This requires bigger dryers and longer retention time.

### Rotary Dryers

Rotary dryers are of two types: single pass and triple pass. The single-pass dryer requires a smaller pressure drop and consumes less fan horsepower. Consequently, it has a lower operating cost. Control of product moisture content is as precise as in a triple-pass dryer; however, this should not be a problem if the end goal of the dryer output is combustion. Both types are available from well-known manufacturers, for example, Koppers, Manufacturing Engineering Construction (MEC), Rader, and Aeroglide among the many vendors with direct-fired or wasteheat-fired single- and triple-pass dryers (Figure 5-11). A single-pass rotary dryer will

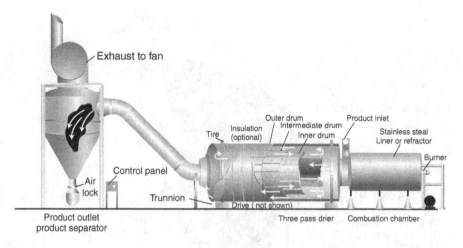

**Figure 5-11.** Rotary wood drying system. (Courtesy Aeroglide Corporation.)

typically process 6 tons/hr of 50% moisture content mill residue in the waste-heat drying mode.

### Cascade Dryers and Flash Dryers

In cascade dryers (Figure 5-12 [3]), wood fuel is dried by falling through streams of hot gas, much like the path of light sand would appear when tossed into a cascading fountain. Cascade dryers tend to be large capacity. No auxiliary fuel input or motor horse-power for rotation are required. However, the pressure drop demand across the unit must be taken up by the fan, so some energy input is necessary. Flash-type dryers are simply several loops of duct in which the wet material and hot flue gas mingle and drying occurs. Successful cascade dryers have been reported from Sweden. The first American installation of a Swedish-designed cascade dryer was completed at a paper mill in upstate New York in mid-1980. The power plant superintendent reported satisfactory operation.

All drying equipment described requires similar auxiliary equipment for material collection and fly ash removal (Figure 5-13). Systems of single cyclones, multiclones, and, where necessary, baghouses, scrubbers, and precipitators, are common. Usually larger wood particles are collected in a precollection hopper, whereas bark fines are separated in cyclones. The dried fuel is conveyed to a storage bin and then into a standard fuel-feed system. Combustibles from the fly ash are collected and mixed with

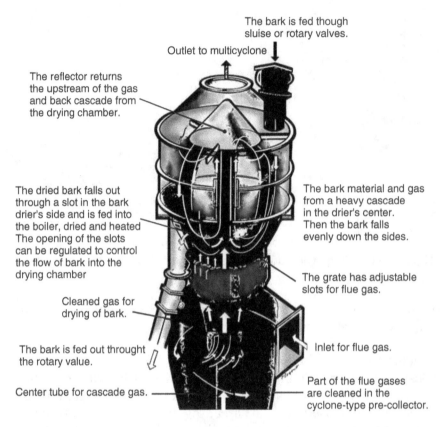

The bark is fed though sluise or rotary valves.

Outlet to multicyclone

The reflector returns the upstream of the gas and back cascade from the drying chamber.

The dried bark falls out through a slot in the bark drier's side and is fed into the boiler, dried and heated The opening of the slots can be regulated to control the flow of bark into the drying chamber

Cleaned gas for drying of bark.

The bark is fed out throught the rotary value.

Center tube for cascade gas.

The bark material and gas from a heavy cascade in the drier's center. Then the bark falls evenly down the sides.

The grate has adjustable slots for flue gas.

Inlet for flue gas.

Part of the flue gases are cleaned in the cyclone-type pre-collector.

**Figure 5-12.** Cascade bark drying system. (Courtesy Bahco Systems, Inc.)

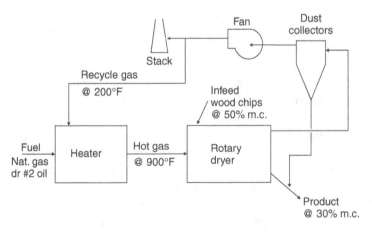

Fan

Dust collectors

Stack

Recycle gas @ 200°F

Infeed wood chips @ 50% m.c.

Fuel Nat. gas dr #2 oil

Heater

Hot gas @ 900°F

Rotary dryer

Product @ 30% m.c.

**Figure 5-13.** Wood drying system flow chart. (Courtesy Stearns-Rogers Manufacturers, Inc.)

the bark fuel, and the remaining fly ash is removed by the pollution control equipment.

## APPROPRIATE APPLICATIONS OF DRYERS

### Retrofits

Retrofitting waste-heat wood-fuel dryers to existing solid-fuel boilers represents the largest market for drying equipment. The reasons for this are economic. Boiler efficiency improvement is effected by decreasing the temperature of exiting flue gas from the neighborhood of 450 to about 200°F out of the dryer. Due to the $SO_2$ acid dew point problem, flue gas exit temperatures of 200°F are feasible only where no sulfur-bearing fossil fuels are being combusted. Where fossil fuel is being combusted in conjunction with wood, flue gas exit temperatures must be maintained at greater than approximately 275°F; the approximate $SO_x$ acid dew point. Indications from the field are that allowing exiting (wood-derived) flue gas to pass below 350°F permits undesirable formation of fouling on ducts, fans, and so on.

As flame temperature increases, less air is moved through the boiler by the fans, and ease of flue gas particulate collection improves due to more complete burning of the fuel in the combustion zone. Often, a plant with steam requirements greater than current operating capacity can gain enough capacity from burning lower moisture content fuel to avoid installation of additional peak-load steam generating equipment. Also, hogged wood or sawdust of greater than 65% moisture content will not normally sustain combustion. In the past, some plants have used fossil-fuel-fired dryers to dry this material so as to avoid storage and/or disposal problems. The advent of dryers fired by waste heat make this wet fuel a potentially attractive fuel source.

In processes in which there are demands for high-temperature off-gases (kilns, textile drying, etc.) increased flame temperatures can be achieved by using dry fuel. The brick industries of Georgia and South Carolina, active in conversion to wood fuels (from natural gas) for firing kilns, have been successful in utilizing waste-heat-fired rotary dryers for obtaining the dry sawdust required for their process, for example, drying wet waste wood from oak flooring operations.

## New Installations

Dryer manufacturers claim that the following benefits are gained when a flue gas dryer is included in the scope of new power plant construction:

1. Reduced dimensions of boiler heat transfer surfaces, air pre-heaters, and economizers
2. Higher efficiency
3. Improved load change response
4. Overall lower costs for new plants incorporating waste-heat wood dryers

## Feasibility for Boiler Steam Plants

Existing solid fuel boilers of at least 60,000 lbs/hr steam capacity are generally the smallest size that should be considered for dryer retrofits. Of course, the larger the boiler and the more expensive the fossil fuel, the more attractive investment returns become. In-stalled costs for dryers are somewhat site specific. A very rough rule of thumb for installed cost is $80,000 per ton/hr of moist fuel input. Payback periods range from one to three years.

## TYPICAL DESIGNS OF WOOD FUEL FACILITIES

The purpose of this section is to present three conceptual designs for wood fuel facilities that address the needs of nonforest-related industrial plants. Ag fuel plants would be similar, mainly differing in feed preparation and handling. The designs can be adapted to meet the requirements for any size plant. Bear in mind that actual designs are site specific and should account for local conditions.

The first system is based on a silo (for storage) and uses pre-pared fuel (fuel that requires no on-site beneficiation). The second system is based on prepared fuel using both outdoor and covered storage. The third system is based on wood waste material and con-tains a fuel preparation section utilizing open and covered storage.

The silo-based system (Figure 5-14 [2]) is a commonly used design for the moderate size plant and for locations in which stor-age area is limited. This system is also indicated where the wood fuel is dry and protection is needed from precipitation. The silo can be made of steel, poured concrete, or staved concrete con-

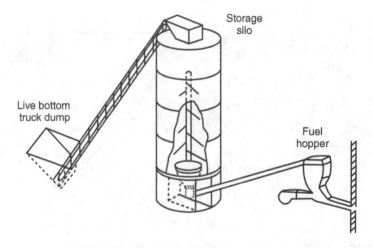

**Figure 5-14.** Silo wood handling system.

struction, and may be almost completely automated. It requires lit-tle attention.

Silos are subject to fuel flow interruptions from bridging and ratholing of the fuel in the silo. For this reason, most silos include a device for breaking up the bridged fuel to assure free flow. There are several devices of this kind. The drawing indicates a frequent-ly used unit that consists of a motor-driven vertical shaft with chain flails. The flow problem also limits the size of the material. Large pieces will not pass through the silo and conveying system, so the wood fuel is limited to pieces 2″ and smaller. These systems are often designed for three days' storage. With this limited stor-age, a reliable fuel supply is imperative.

The next system (Figure 5-15 [2]) consists of open and cov-ered storage with wood handling done by a front-end loader. For a given size, this design represents the lowest capital cost. No con-veying equipment is specified, though for a particular installation, conveying equipment may be desirable.

Fuel that does not bridge in normal operation may well do so when left idle in the silo for too long. Operators have found this to be the case during a three day shutdown. The fix is simple: Pro-vide a way to withdraw material and divert it from the boiler or combustor at least once per shift. This can be returned to the top of the silo to keep the material form compacting and bridging.

The open storage consists of a concrete slab designed to hold a 30 day supply of wood fuel. The slab has a slight pitch to aid in

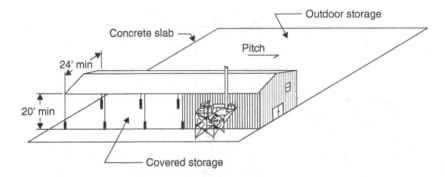

**Figure 5-15.** Outdoor and covered storage.

rainwater runoff. The covered storage is designed to hold three days of fuel and consists of a concrete slab with an open-sided metal building. In practice, the covered storage is the active area, with the outdoor storage serving as a reserve holding area to allow for special purchases and to provide fuel during periods when the regular supply may be interrupted.

No fuel preparation is included, as it is anticipated that the fuel will be purchased already sized to the requirements of the

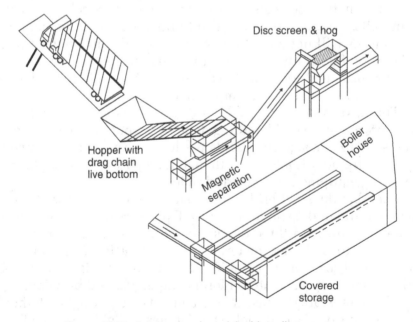

**Figure 5-16.** Large wood yard fuel handling systems.

combustion system. A truck receiving–unloading system may have to be included, depending upon the type of fuel delivery.

The last system (Figure 5-16 [2]) illustrates the components that are usually necessary for the larger systems. The fuel is received from a hydraulic truck dump and placed into a live-bottom pit. The material is then conveyed up through a station where it is passed over a magnetic separation belt to remove tramp metal. Next, the fuel is conveyed to a disc screen and hog where any oversized pieces are reduced to proper size. From the hog, the material is transported by conveyors to the covered storage. (Not shown but generally included in a system of this size is an open storage area.) After passing through the disc screen and hog, the conveying system is so arranged that the fuel can be transported to covered storage or to an open storage area. A front-end loader is a necessary piece of equipment to handle special transport problems and to handle spills.

## REFERENCES

[1] Curtis, M. L, "The Effects of Outside Storage on the Fuel Potential of Green Residues," 1980, M.S. Thesis, Virginia Polytechnic Institute and State University, Blacksburg, VA.

[2] Curtis, G. B., Brown, M. L., and others, "Wood Energy Economics— Proceedings April 30, 1980," Georgia Institute of Technology, Engineering Experiment Station, April 1980.

[3] Curtis, G. B., "Hazards Associated with Industrial Wood Combustion," Georgia Institute of Technology, Engineering Experiment Station, Atlanta, GA, 1979.

CHAPTER **6**

# COGENERATION AND POWER GENERATION

## INTRODUCTION

Cogeneration can be defined as the simultaneous production of electricity and thermal energy. In the case of a wood boiler, the thermal energy will be in the form of process steam. An industrial plant planning a major expansion of steam producing capabilities would do well to examine the cogeneration option.

Hardware is available to industrial energy users for generation of their own power, and the source of energy for the steam turbines could well be a biomass-fired boiler system. A second option has been across-the-fence-line sale of steam from stand-alone independent power production.

The mechanics of working out cooperative agreements with local power companies is always an issue. Smaller power producers may opt to produce less power than they consume, and avoid the issue of power sales. Large pulp and paper mills in the Southeast generate all or part of their electricity (largely supplied by wood-fired boilers) and some have sold power to utility grids, but that situation is not widespread. Utility companies usually cite these major drawbacks: (1) industrially generated power is often of low quality (due to inadequate frequency and voltage control) and (2) the industrial producer often produces excess power at off-peak times of the day when it is not needed by the utility.

## GOVERNMENT INCENTIVES

An impediment to the wide acceptance of cogeneration has been the unanswered question of cogenerator/utility interface. Often, potential cogenerators were discouraged because utilities were under no obligation to buy excess power. When electricity was offered for sale, the cogenerators could then be subject to the same strict laws as a utility. In March 1980, however, federal legislation was enacted to encourage further cogeneration ventures.

These rules state that utilities must purchase electricity from cogenerators at the utilities' "avoided" cost. An avoided cost is that amount a utility saves in fuel and capacity additions costs.

## STEAM TURBINES

The prime mover used for cogeneration of process steam and electricity is a steam turbine. Steam turbines can be grouped very broadly into condensing or noncondensing units. Since steam exhaust is condensed at the exhaust of a condensing turbine, process steam at the desired pressure must be extracted before the exhaust. Extraction valves are then included to supply the process steam. In the noncondensing case, the turbine produces power by acting as a pressure reducer and the turbine exhaust becomes process steam.

The selection of a condensing or noncondensing turbine is based on such factors as boiler pressure, process steam requirements, electrical requirements, and cooling water availability. Steam turbine types suitable for industrial cogeneration are listed in Table 6-1. If the most important boiler product is process steam, a noncondensing turbine is generally specified. Operation requires increasing the boiler pressure that is used in the turbine to produce power. In the noncondensing unit, process steam pressure and flow are requirements to be satisfied. Therefore, since electric generation is a direct function of steam flow, periods of low steam demand correspond to low electrical generation. Noncondensing turbines cannot meet all variations in electrical load and an alternate power source must be provided.

If the turbine–generator set is expected to meet all variations in both process steam and electrical load, a condensing turbine is necessary. Power is generated by steam flowing through the turbine and process steam is provided by extraction. In times of high steam demand, most of the power is generated by the steam before it is extracted. At other times, the entire steam flow is through the turbine to the condenser and electricity is produced. This arrange-

**Table 6-1.** Turbine–generator configurations

| Turbine type | Application |
|---|---|
| Back pressure (noncondensing) | Turbine exhaust meets process steam demand; electricity generation directly related to steam flow |
| Single automatic extraction (condensing) | Supplies process steam at one pressure level, meets variations in electrical load |
| Single automatic extraction (noncondensing) | Supplies process steam at two pressure levels, electrical generation directly related to steam flow |
| Double automatic extraction (condensing) | Supplies process steam at two pressure levels, meets variations in electrical load |
| Double automatic extraction (noncondensing) | Supplies process steam at three pressure levels, electrical generation directly related to steam flow |

ment allows variation of process steam and electrical production independently. Three disadvantages of condensing turbines are system size limitations, high cost, and overall cycle efficiency. Condensing turbines are difficult to produce in the smaller power ranges and sizes below 5000 kW are generally not available. Overall system cost is increased with the addition of a condenser, piping, and a circulating water system. Overall cycle efficiency with a condensing unit is reduced as virtually all the usable heat contained in the exhaust steam is lost to the cooling medium in the condenser.

Manufacturers were surveyed to get representative cost figures for four different sized cogeneration systems. Results for 1, 5, 10, and 25 MW installations are presented in Table 6-2. Installations include turbine, generator, baseplate, and lube systems. Piping, switchgear, condenser, and other auxiliary elements are not included. Although there are variations, some general conclusions can be reached. For a one megawatt installation, the turbine suitable is a single-stage noncondensing unit. In the power ranges above this, either a condensing or noncondensing unit can be used. Most manufacturers specified a noncondensing unit, although one was quoted both ways. Turbine maintenance costs varied widely, but all were calculated one of two ways. Annual maintenance costs are either estimated as a percentage of capital investment (2 to 2½%) or as a multiplier times power production in cents/kWh.

## ECONOMIC CONSIDERATIONS

The alternatives to be considered when deciding upon a cogeneration system are:

- Purchase power with separate steam generation
- Cogeneration of power and steam
    Back-pressure turbine
    Condensing turbine

Factors that influence the decision include purchased electrical cost, fuel cost, and required investment ($/kW). General cost figures for turbine generator equipment are shown in Table 6-3. Note the increased cost for the condensing turbine option. As an illustration, consider a plant with the following requirements:

- Steam flow          101,500 lb/hr
- Steam pressure      250 psig
- Steam temperature   445°F

Figure 6-1 [1] is a typical performance map for a steam turbine. This performance map is used to illustrate the output of the turbine supplying process steam at the conditions given above.

**Table 6-2.** Turbine price estimates

| Manufacturer | A | B | C | D | E | Comments |
|---|---|---|---|---|---|---|
| 1 MW | $1,050,000 | $260,000 | $630,000 | $270,000 | $320,000 | Single stage |
| | (1050 $/kW) | (260 $/kW) | (630 $/kW) | (270 $/kW) | (320 $/kW) | Noncondensing |
| Maintenance, | $26,000 | $15,000 | | | | |
| annual | $2,500,000 | $1,620,000 | $1,600,000 | $1,300,000 | $1,300,000 | Multistage |
| 5 MW | (500 $/kW) | (320 $/kW) | (320 $/kW) | (250 $/kW) | (250 $/kW) | condensing or |
| | | | | | | noncondensing |
| Maintenance, | $62,000 | $74,000 | | $25,000 | | |
| annual | $3,300,000 | $2,500,000 | $2,400,000 | | $2,300,000 | Multistage |
| 10 MW | (330 $/kW) | (250 $/kW) | (240 $/kW) | | | condensing or |
| | | | | | | noncondensing |
| Maintenance, | $830,000 | $130,000 | | | | |
| annual | $7,200,000 | $5,300,000 | | | $7,100,000 | Multistage |
| 25 MW | (290 $/kW) | (210 $/kW) | | | | condensing or |
| | | | | | | noncondensing |
| Maintenance, | $180,000 | $400,000 | | | | |
| annual | | | | | | |

Note: Turbine prices escalated via 2.1 CPI inflation factor, 1984 to 2008.

**Table 6-3.** Turbine costs: condensing versus noncondensing

|  | Noncondensing | Condensing |
|---|---|---|
| Turbine–generator | $760,000 | $1,500,000 |
| Condenser | — | $210,000 |
| Cooling tower | — | $190,000 |
| Total | $760,000 ($300/kW) | $1,900,00 ($370/kW) |

Note: 1. Does not include piping, insulation, instrumentation, electrical, or switchgear.
2. Turbine cost escalated via 2.1 CPI inflation factor, 1984 to 2008.

The line A-F-B shows all the possible operating conditions for the back-pressure case. Electrical output is directly proportional to steam flow, and turbine rated output is reached at rated process steam flow. This line would also be the operating line for the condensing case when all the steam is being extracted and none flows to the condenser. Using the condenser for control allows the condensing turbine to operate anywhere in the area bounded by A-B-E-C. The performance map illustrates that power is produced both by the extraction flow and flow in the condenser. To supply the

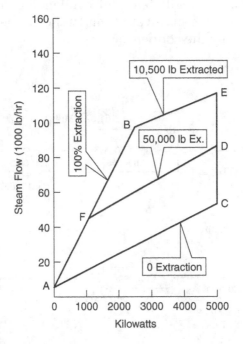

**Figure 6-1.** Steam turbine performance map.

turbine, the boiler pressure is increased to 650 psig in both the condensing and noncondensing cases, but the steam flow is increased only for the condensing case.

Figures 6-2 through 6-6 [1] show the operating conditions for the boiler alone and with four different turbine options. The heat balances for each case are shown in Figures 6-3 through 6-6. The boiler-only case shows fuel consumption, increased fuel consumption over the boiler-only case, and the electrical cost using fuel at $1.65/MMBtu ($15/ton) for four turbine cases. For the back-pressure case, electricity is generated by steam required for the process. Two extraction cases are examined: (1) Case A, in which half the power is generated by process steam before extraction and half is generated by steam flow to the condensers; and (2) Case B, in which one-third is generated by extraction steam and two-thirds by condenser steam flow. The final case, that of the condensing turbine, is shown as a limiting case. Notice that electrical costs (Table 6-4) are higher for extraction turbines than for back-pressure turbines. The electrical cost increases in proportion to the amount of generation by condenser flow and reaches the limit in the condensing-only turbine. The difference in electrical cost is due to the relative inefficiency of condensing versus back-pressure turbines in this application and reflects the cost of control being provided by the condenser.

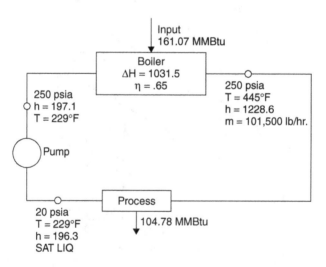

**Figure 6-2.** Wood-fired boiler cycle.

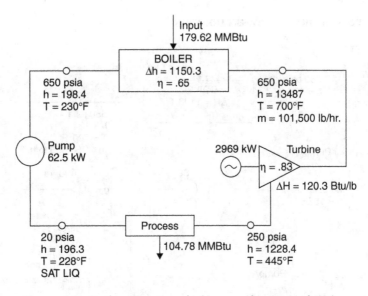

**Figure 6-3.** Back-pressure cycle (noncondensing turbine).

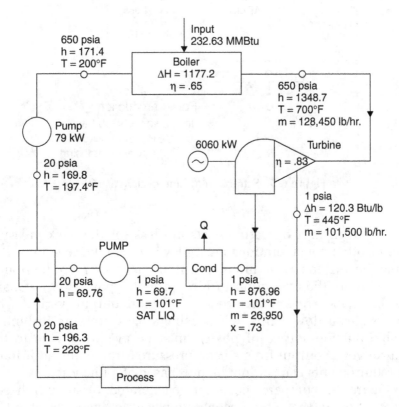

**Figure 6-4.** Extraction turbine cycle (case A).

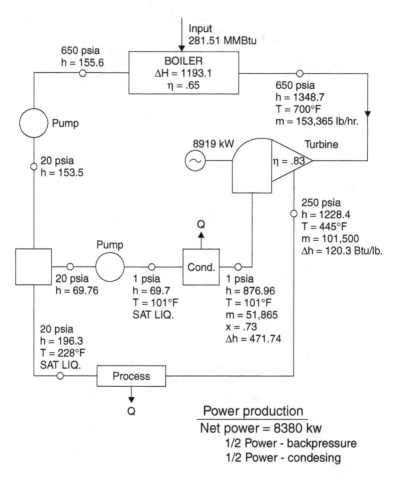

**Figure 6-5.** Extraction turbine cycle (case B).

Although unit operating costs are less for the back-pressure turbine, other considerations must play a role in the selection. The extraction turbine costs more to install, yet power production is greater. Also, the extraction turbine provides power on a consistent basis, whereas back-pressure turbine output varies with process steam flow. Thus, when using a back-pressure turbine, a reliable outside source of power must be available. Due to the wide surges in output from a back-pressure turbine, the additional kW requirements can seldom be purchased at a low rate.

Figure 6-7 compares costs for cogenerated versus purchased power. Notice that the cogeneration curve ends at about 1

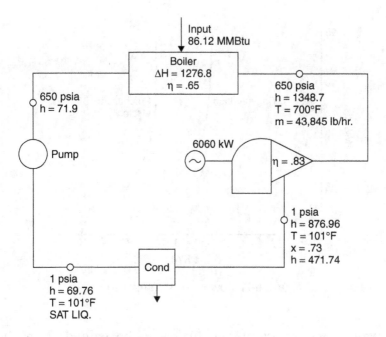

**Figure 6-6.** Condensing cycle (no extraction).

**Table 6-4.** Turbine operating conditions and electrical production costs

|  | Boiler only | Back pressure | Extraction (A) | Extraction (B) | Condensing |
|---|---|---|---|---|---|
| Boiler pressure (psig) | 250.00 | 650.00 | 650.00 | 650.00 | 650.00 |
| Boiler temperature (°F) | 445.00 | 700.00 | 700.00 | 700.00 | 700.00 |
| Steam flow (lb/hr) | 101,500 | 101,500 | 128,450 | 153,365 | 43,850 |
| Electrical output (kW) | — | 2,900.00 | 5,700.00 | 8,380.00 | 5,700.00 |
| Fuel consumption (MM Btu/hr) | 161.07 | 179.62 | 232.63 | 281.51 | 86.12 |
| Increased consumption (MM Btu/hr) | 0 | 18.55 | 71.56 | 120.44 | — |
| Electrical cost (g/kW)* | — | 1.05[†] | 2.07[†] | 2.37[†] | 5.20[‡] |

*Based on $15/ton wood fuel.
[†]Based on incremental fuel use for power.
[‡]Based on total fuel used.

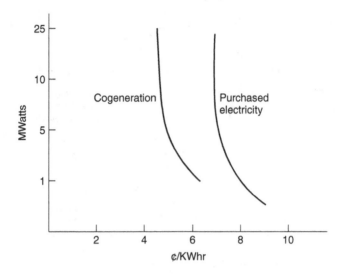

**Figure 6-7.** Power costs comparison.

megawatt. The important conclusion to be gained from these curves is that for the plant considering purchasing a new boiler in the 30,000 to 50,000 lb-steam/hr range, the additional cost of cogeneration equipment becomes highly justifiable.

## COGENERATION SYSTEM SIZE

Cogeneration systems in large sizes (above 20 MW) have been in wide use for many years; however, rising utility costs have resulted in many smaller scale users examining this option. Several disadvantages tend not to favor electrical production from boilers smaller than 50,000 lb/hr, roughly equivalent to 5 MW output. The efficiency is low for this size of steam turbine, with steam rates as high as 60 lb/kW hr, meaning decreased electrical production. Table 6-5 shows the expected power production from a back-pressure turbine for four different steam flow rates. Also, because of their size, no economy of scale is realized with steam turbines this small and capital cost can be roughly three times as much per kilowatt as large steam turbines. The electrical production can be increased if a condensing unit is utilized. Use of a condensing turbine means that little or no steam will be available for process applications, and this is unacceptable for most industrial situations

**Table 6-5.** Electrical production for various steam flow rates of back pressure turbine

| Steam flow (lb/hr) | Approximate steam rate (kW/lb/hr) | Electrical production (kW) |
|---|---|---|
| 10,000 | 55.5 | 180 |
| 25,000 | 46.3 | 540 |
| 50,000 | 41.7 | 1,200 |
| 100,000 | 40.0 | 2,500 |

since process steam is the primary purpose for having a boiler. In summary, electrical generators for cogeneration systems can be purchased in all sizes; however, economics tend to favor turbines of 5 MW and larger.

## REFERENCE

[1] Committee on Industrial Ventilation, *Industrial Ventilation—A Manual of Recommended Practice,* American Conference of Government Hygienists, Lansing, MI, 1978.

# CHAPTER 7

# EMISSIONS AND CONTROL

## INTRODUCTION

The primary environmental cause for concern in the burning of biomass is the emission of particulate matter. The presence of smoke indicates the presence of particulates, but the relationship between smoke and particulates is not easily quantified. Smoke results primarily from a combination of inorganic ash, noncombustibles, particles of carbon, small liquid aerosols, and other combustible matter that have not burned completely. For wood combustion, the emission of most concern is particulate matter related to unburned carbon and the ash content of wood. A dark plume from an industrial stack can indicate poor maintenance and poor operational practice, or an unavoidable situation such as a rapid load change, boiler upset, or soot blowing.

The 1976 Clean Air Act and later amendments established rules and regulations that govern the emission of certain materials into the atmosphere. As it is an expense that can be substantial, both in initial and operating costs, emission control deserves careful attention from the system designer.

Emission control devices cannot be considered in isolation from the other parts of the system. Fuel, burner, and boiler design all influence the particulate emission rate. For example, with low-turbulence combustion, more of the ash remains in the combustion chamber and does not pass through the boiler. The emphasis must be on system design to provide an installation that will comply with regulation.

*Biomass and Alternate Fuel Systems.* Edited by McGowan, Brown, Bulpitt, Walsh
Copyright © 2009 American Institute of Chemical Engineers, Inc.

## TYPES OF EMISSION

Particulates from the combustion of wood are composed of ash, unburned carbon, condensed droplets, and sand and other silicates foreign to the fuel itself. As do other combustion sources, wood furnaces emit carbon monoxide (CO), oxides of nitrogen ($NO_x$), trace amounts of oxides of sulfur ($SO_2$), and unburned hydrocarbons (HC). These generally are either too low to be of concern or are not regulated by law. Because wood is lower in ash content than coal, it tends to produce less particulates and has almost no sulfur oxides when compared with coal. $NO_x$ is generally low due to lower combustion temperatures (compared to coal) and little or no fuel-bound nitrogen. The exception would be when burning dry wood or ag residue at low excess air conditions; flame temperatures would then approach that of fossil fuels. From an emissions viewpoint, biomass can be an attractive alternative.

## EMISSION REGULATIONS

Applicable air pollution standards for emissions vary from state to state. In Georgia, the emission criteria to be considered by prospective operators of wood-fired boilers is included in "Rules and Regulations of Air Quality Control," published by the Georgia Department of Natural Resources (available at www.gaepd.org/Documents/index_air.html). Figure 7-1 illustrates the maximum permissible particulate emissions from fuel burning installations. Allowable emissions are based on total input in millions of Btus per hour, beginning on the horizontal scale with 1 million Btus per hour. Varying from state to state, the applicable regulations should be determined for each location, and may be affected by local air quality (e.g., being more stringent in nonattainment areas) and type of application. For example, subparts of Georgia Rule 391-3-1-.02(2) apply to wood burning facilities, with extra requirements for electricity generating facilities over 25 MW in size, and that cofire coal at least part of the time. The Georgia rule also varies the $NO_x$ requirements (25 vs. 100 tpy) based on county location tied to nonattainment status.

"Percent plume opacity" refers to the amount of light blocked by the source plume. An opacity of 100% will theoretically allow no light to pass through, whereas the background will only be obscured 20% by a plume of 20% opacity. Personnel who have been trained in the EPA method (sometimes called "smoke readers") do opacity readings visually. The State of Georgia has a general opacity limit of 20%, except for six one minute periods within a given hour when 27% opacity is allowed to permit soot blowing or system upsets. Wood-fired units of moderate to large size will be hard pressed to meet opacity standards and the accompanying particulate regulation without some form of control equipment. Completion of smoke reading schools conducted by the Georgia Department of Natural Resources (DNR)—three-day sessions held several times per year—will result in the attendee becoming a Certified Smoke Reader.

Companies buying a boiler without help from a design firm should be aware of the appropriate emission control standards. Before constructing a wood-fired boiler in Georgia, one should submit form APCS-APC-2 (an application to construct fuel burning equipment) to the Air Quality Control section of the Department of Natural Resources. This form is reviewed, and if in the agency's judgment the source will comply with existing emissions control standards, it is approved.

The critical path on which the start up of new combustion equipment travels often involves crossing a bridge of lengthy paperwork. For this reason, it is recommended that air pollution regulatory authorities be involved in the project from the earliest possible stage, thereby minimizing risks of expensive and time-consuming installation redesign.

There is another regulation that may apply depending on the outcome of a review. To maintain air quality at the current levels, the U.S. EPA promulgated regulations [to prevent significant deterioration (PSD) of air] for sources that emit above 250 tpy (tons per year) of fly ash. Generally, for the majority of small industrial boilers, PSD regulations will not apply. All sources not regulated by PSD standards will fall under the existing state particulate and opacity rules already presented. An EPA workshop manual, "Prevention of Significant Deterioration," may be of interest [1].

After construction, the owner must file for a permit to operate. If no significant design changes have been made since filing for a permit to construct, a short form to operate may be sufficient. Many of the application forms are available online from Georgia EPD.

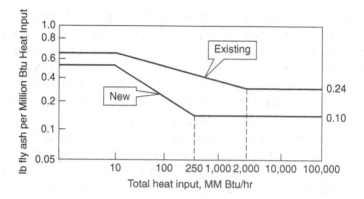

**Figure 7-1.** Maximum permissible emissions of fly ash and particulates. (State of Georgia.)

## EXPECTED EMISSIONS

### Areas Affecting Emissions

Combustion and the formation of emissions is a complicated process controlled by numerous variables. The three major areas that affect biomass-fired boiler emissions are fuel, boiler design, and boiler operation. Text on boiler emissions would apply too to hot oil heaters and hot gas heaters using grate firing. Emissions from hot gas heaters based on fluid beds are addressed separately.

### Fuel

The importance of fuel size and size consistency cannot be overemphasized. Oversize pieces are difficult to distribute evenly in the furnace and tend to burn slowly. When suspension burning occurs, fine particles are easily entrained in the flue gases, leading to higher particulate loading. The size of the fuel should be consistent with equipment design parameters, aiding distribution on the bed and assuring equitable distribution of the underfire air. The cleanliness of the fuel must be considered in terms of whatever foreign matter is picked up during harvest and transport and then fed into the boiler. Ash content of wood itself is low; but material lodged in the bark during removal from the forest, or scooped up by the loader during handling, effectively increases the noncombustible fraction of the wood fuel. Many ag fuels may have more ash than wood. Burning high ash content papermill sludge in bark boilers will add to ash levels and could raise PM emissions.

## Boiler Design

Wood-burning boilers are designed to maximize combustion efficiency and limit emissions. Designers strive to minimize the gas velocity in the furnace by increasing the grate area. A large grate area decreases the air flow per unit, resulting in lower velocities. Lower velocities help reduce emissions by increasing the fuel residence time in the combustion zone and decreasing entrainment of fine particles by high speed air. Another factor that affects emissions is the method by which fuel is introduced into the boiler. Spreader stokers will have some suspension burning and generally higher emissions than methods that burn primarily on the grate. Particles burning in suspension are easily carried out of the furnace into the backpasses of the boiler. It is a common experience with wood burning that a percentage of the fuel may not completely combust and be carried out of the boiler unburned. This will be caught in the primary collector; frequently, this is a multiclone mechanical collector. Especially in large size boilers, the catch from the multiclone is reinjected into the furnace to burn more of the carbon. Though this practice improves boiler efficiency, reinjecting char tends to increase the percentage of small particles. Figure 7-2 illustrates the effects of reinjection on the particle loading. Note that 100% reinjection is not achievable since this would involve injecting already collected fly ash fines. For some combustors, char is reinjected from the first-stage multiclone and is disposed of from a second-stage multiclone.

## Boiler Operation

Boiler operating methods and firing techniques have a direct effect on stack emissions. Unlike most natural gas or light-fuel-oil-fired boilers, wood-fueled installations require full-time operator attention to achieve maximum efficiency. For example, the rate of flow of combustion air can be automated to the flow of natural gas or oil fuel, but this is not practical with wood systems. If the boiler fireman consistently supplies less air than is needed, unburned fuel will result and smoke will be produced. Excess combustion air will reduce combustion efficiency and increase the total mass of combustion products. This increased mass results in higher velocities in the combustion zone, which causes the entrainment of particulate matter and reduces the fuel residence time.

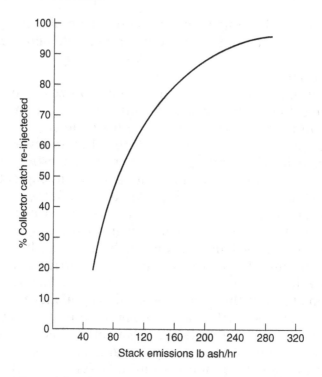

**Figure 7-2.** Effect of fly ash reinjection on particulate emissions. (From A. Barron, "Studies On The Collection Of Bark Char Throughout The Industry," *Tappi Journal,* Vol. 53, August 1970.)

In addition to the quantity of combustion air, the method of its introduction can influence emissions. Wood combustion systems are often designed to introduce a percentage of the combustion air from beneath the grates and the remainder from above the grate line (under- and overfire air). The underfire air is necessary for combusting the fuel on the grates, whereas the overfire air is used to combust the fuel volatile gases. Although the ratio of over- to underfire air is a function of boiler design, a good fireman adjusts the air to meet the needs of particular fuels and operating conditions.

When a boiler is overloaded, a condition similar to that of high excess air develops. As the rated load is exceeded, air flow to the boiler is increased, leading to higher velocities and more particle entrainment. Higher-than-normal emissions can be expected when a boiler is operated over its rating.

Upsets in the boiler operation can cause temporary excess emissions. Upsets can occur when the load changes rapidly or when the moisture content of the fuel varies unexpectedly. Though upsets with wood-fired boilers are difficult to eliminate, proper corrective action by the operator can reduce their effect.

## Fluidized-Bed Hot Gas Heaters

Fluidized-bed combustors have relatively high gas velocities leaving the bed. Larger particles stay in the bed, but as they grow smaller, they are entrained in the gas stream. Traditionally, one or to stages of cyclones capture the particles. The first stage may be reinjected into the bed to reburn carbon, whereas the second stage may exit the system and go to disposal. Fluid beds frequently have a bed drain for oversize material in addition to fly ash treatment.

## Suspension Burning

Burning dry planer shavings or sanding dust in a cyclonic or high-swirl burner will produce fine particulates, likely to be low in carbon and high in ash.

## Emission Factors

Emission factors for particulates emitted from wood-waste boilers without clean-up systems are presented in Table 7-1. The factors are presented as pounds of particulates per ton of fuel burned. The factors for bark and bark mixtures are based on an as-fired moisture content of 50%. The wood fuel considered includes sawdust (5 to 50% moisture), shavings, ends, and so on, but no bark. For well-designed and operated boilers, use lower emission factors; and in the opposite case, use higher values. The emission factor for wood only is expressed on an as-fired moisture content basis assuming no fly ash reinjection.

The U.S. Environmental Protection Agency (EPA) compiles expected emission levels from combustion sources based on pounds of fly ash per ton of fuel burned. These factors can only be used for approximate calculations because they do not consider the effects of variables mentioned earlier.

## Emission Characteristics

Several properties of particulate matter from biomass combustion are important when considering control equipment. The particle

**Table 7-1.** Particulate emission factors for wood and bark combustion in boilers*

| Fuel | Emission (lb/ton of fuel as fired) |
|---|---|
| Bark: | |
| with reinjection | 75 |
| without reinjection | 50 |
| Wood/bark mixtures: | |
| with reinjection | 45 |
| without reinjection | 30 |
| Wood | 5–15 |

*From AP-42, "Compilation of Air Pollutant Emission Factors," U.S. Environmental Protection Agency, 1975.

size distribution is an important factor in control device operation. Smaller particles (less than 10 microns) are difficult to trap. Gas streams containing large percentages of submicron (less than 1 micron) particles require robust equipment for removal, and venturi scrubbers (an old technology) require large energy input to do the job and are no longer a good choice when faced with current regulatory limits. Fortunately, particulates from wood combustion are relatively large in size (Figure 7-3).

Particle strength is also a property to be considered in collection. Carbon particles break very easily into smaller particles, making collection by mechanical means difficult. Therefore, the first-stage mechanical collectors used to capture unburned char are designed for relatively low velocities to limit further attrition.

Particulate resistivity is an important property to consider if a dry electrostatic precipitator (dry ESP) is to be used for collection. Particles must exhibit a relatively high resistivity (the ability to accept and hold a charge) to be collected effectively. Wood ash generally has a low resistivity, making it more difficult to collect by electrostatic precipitators. Wet ESPs are not affected by particle resistivity.

## CONTROL TECHNIQUES

Many devices are currently available for the control of emissions from combustion sources. The five major control devices in use today are mechanical collectors, baghouses, wet scrubbers, dry

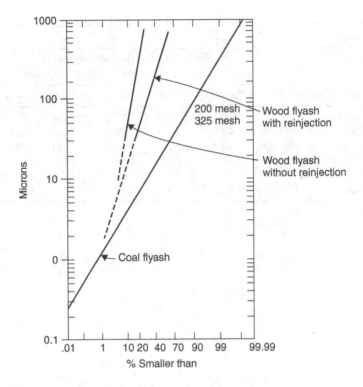

**Figure 7-3.** Size distribution of wood and coal fly ash. (From "Comparison Of Fossil And Wood Fuels," U.S. Environment Protection Agency Report, EPA-60012-76-256, March 1976.)

scrubbers, and wet or dry electrostatic precipitators. Table 7-2 shows the size range of particulates each device will collect efficiently and economically.

## Mechanical Collectors

The most common control device is a cyclone, which is a centrifugal dust collector. Cyclones are used primarily to collect larger dust particles. In the period before the emphasis on emission control, their purpose was to clean the airstream of large particles in order to protect induced-draft fans against erosion and allow refiring of high-carbon ash. Particles are removed through the action of a double vortex. The inlet gases spiral downward, creating a centrifugal force that pushes the particulates toward the wall (Figure 7-4) [4]. The particulates drop out as the gases change direction

**Table 7-2.** Relative collection efficiencies of various pollution control devices [3]

| Type of collector | Collection efficiency, % for particle size range, microns | | | | | |
|---|---|---|---|---|---|---|
| | Overall | 0–5 | 5–10 | 10–20 | 20–44 | 44 |
| Simple collector | 65.3 | 12.0 | 33.0 | 57.0 | 82.0 | 91.0 |
| Multiple cyclone (12″ dia) | 74.2 | 25.0 | 54.0 | 74.0 | 95.0 | 98.0 |
| Multiple cyclone (6″ dia) | 93.8 | 63.0 | 95.0 | 98.0 | 99.5 | 100 |
| Electrostatic precipitator | 97.0 | 72.0 | 94.5 | 97.0 | 99.5 | 100 |
| Venturi scrubber | 99.5 | 99.0 | 99.5 | 100 | 100 | 100 |
| Baghouse | 99.7 | 99.5 | 100 | 100 | 100 | 100 |

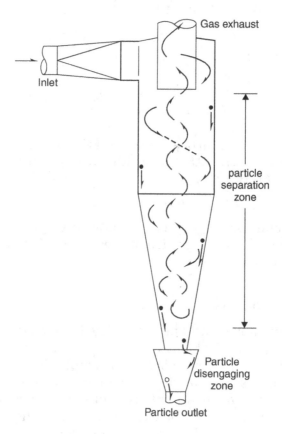

**Figure 7-4.** Cyclone collector.

and spiral upward to the exit. Cyclone collectors are best at removing large particles; their efficiency decreases as the mean particle size decreases. An important controlling parameter is the cyclone diameter. As the diameter increases, particles must travel farther to reach the wall; thus, efficiency decreases. This diameter limitation is overcome by using a multiclone (multicyclone) which is a series of small-diameter cyclones arranged for parallel flow. Multiclones can handle the same airflow as large diameter single cyclones but at a higher efficiency. The expected pressure drop through a cyclone or multiclone is 1" to 6" of a water column. If a mechanical cyclone collector is insufficient to meet emission standards, it is usually used as the first stage of collection, followed by an appropriate secondary collector.

## Baghouse

A baghouse functions like a huge vacuum cleaner with multiple bags stretched over wire cages (Figure 7-5 [4]). The most popular type is the pulse jet, with air entering from the bottom, flowing through the bags, and out the top. As air passes through the bags, particulate matter is trapped and forms a dust cake, which does the filtering. The trapped particles are removed from the bags by a periodic brief reverse pulse of high-pressure air. Other types of baghouses exist, and very large ones (e.g., those used for pulp mill boilers) may be of the reverse-air type. These have the dirty gas on the inside of long, large-diameter bags, and isolate a compartment, then reverse the airflow to collapse and clean the bags. Baghouses are extremely efficient (above 99%, even for submicron particles), have a low pressure drop, and require low fan horsepower.

Historically, baghouses have seen little application in wood-fired boiler applications because of the fire hazard created by spark carryover. Bags typically are replaced every 18 to 24 months. The bags are temperature limited to 500°F operating temperature, with excursions to 550°F. Among the bag materials in this temperature range are fiberglass with Teflon coating, Nomex, felted Teflon, and P84. A mechanical collector is used upstream from a baghouse to remove large glowing particles, and spark arrestors may also be used. Operating the baghouse at positive pressure to prevent air leakage into the baghouse reduces the possibility of fires by keeping the baghouse atmosphere in a low oxygen

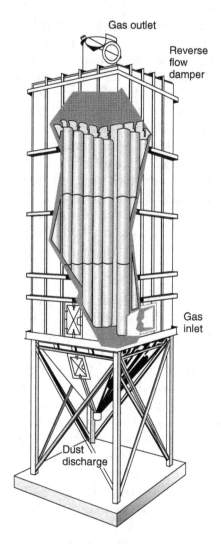

**Figure 7-5.** Reverse air fabric filter system. (Courtesy Zurn Industries Inc.)

condition and reduces the chance of oxygen causing any glowing char to burn at a significant rate.

### Wet Scrubbers

Wet scrubbers are designed to develop an interface between a scrubbing liquid and the gas to be cleaned. The particles in the gas are trapped by liquid droplets; then the liquid is collected and removed. There are a variety of wet scrubber designs available: baffle and spray (Figure 7-6 [4]), venturi (Figure 7-7 [4]), and impingement (Figure 7-8 [4]) to name a few. The type of wet scrubber is chosen based on allowable emissions, particulate size distribution, and gas conditions going into the scrubber. In wood applications, baffle-and-spray or venturi types are usually used.

Wet scrubbers have the advantage of high efficiency (even on small particles), relatively low first and maintenance costs, and resistance to fire damage since glowing sparks are water-quenched. The obvious disadvantage of a wet scrubber is the problem of water supply and sludge disposal. Wet scrubbers have a high demand for water (usually 5 to 10 gallons per ACFM (actual cubic feet per minute) for evaporation and blowdown). Once the liquid has collected the particles, it must be cleaned through the use of a clarifier or settling pond. Also, the energy requirements of a wet

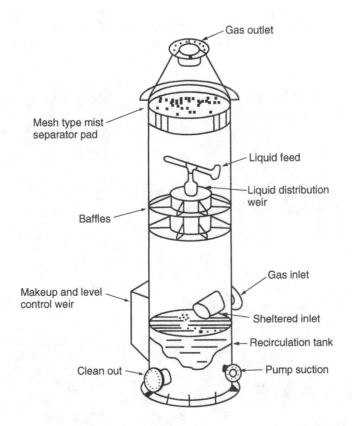

**Figure 7-6.** Baffle and spray. (Courtesy of Andersen 2000 Inc.)

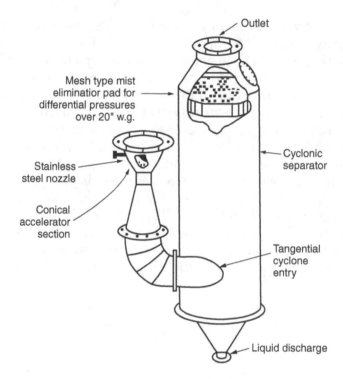

**Figure 7-7.** Venturi, (Courtesy of Andersen 2000 Inc.)

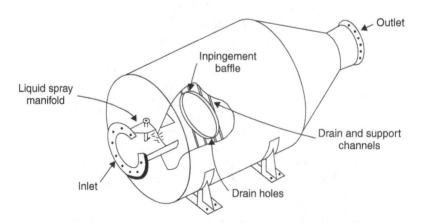

**Figure 7-8.** Impingement scrubber. (Courtesy of Andersen 2000 Inc.)

scrubber increase as mean particle size to be collected decreases. Traditionally, wood applications require a low-to-medium-energy scrubber (6″–15″ water column), but this will not get stack particulate matter (PM) low enough to meet modern emissions limits for PM.

### Electrostatic Precipitators

Electrostatic precipitators (ESPs) (Figure 7-9 [4]) operate by ionizing particles as they enter the devices. These charged particles are then attracted to oppositely charged collection plates. The plates are cleaned by periodical mechanical "rapping." Dislodged particles are collected in a hopper. The collection efficiency strongly depends on the resistivity of the particles to be collected. If the

**Figure 7-9.** Large electrostatic precipitator. (Reprinted with the permission of Environmental Products Division, Dresser Industries, Inc.)

electrical resistivity of the particles is extremely high or low, the efficiency of an electrostatic precipitator will be adversely affected. Because particulate emissions from wood boilers are generally low in resistivity, collection efficiency suffers in dry electrostatic precipitators. Reentrainment is another problem. It occurs when plates are rapped, dust falls into the hopper, and some dust goes back into the airstream and out the stack. Use of two-stage precipitators reduces this problem, and some plants that prefer a dry fly-ash product do use dry ESPs.

Wet ESPs (WESPs) have been used in a wide variety of PM removal applications. Ceilcote and Research Cottrell are two major names in the business. They are very effective at removing small particles. The Ceilcote type has a brief charging/collection section followed by a packed bed for further particle collection. They are usually used with two (or more) modules in series to combat loss of collection efficiency when power is turned off and plates are washed. Both types have low pressure drops and moderate overall power requirements. If biomass plus coal is burned (or other sulfur-containing fuel), they can be used for $SO_2$ removal also.

## COSTS

Estimated costs per cubic foot per minute (CFM) of flue gas for various emission control systems are presented in Table 7-3. These are approximate installed costs for the aboveground structure. These costs are shown as a comparison between various con-

**Table 7-3.** Pollution control equipment installation and operating costs [2]

|  | Installed cost ($/cfm) | $\Delta P$ (in $H_2O$) | Yearly energy cost for fan operation (dollars) |
|---|---|---|---|
| Single multicyclone | 2 | 2 | 8,000 |
| Double multicyclone | 4 | 4 | 17,000 |
| Spray and baffle wet scrubber | 4 | 1–3 | 5,900 |
| Venturi wet scrubber | 6 | 10 | 21,000 |
| Dry scrubber | 8 | 6 | 12,000 |
| Baghouse | 11 | 6 | 12,000 |
| Electrostatic precipitator | 15 | 0.5 | 2,000 |

Note: Costs inflated 1984 to 2008 via CPI factor of 2.1.

trol methods and should not be regarded as final, since site-specif-
ic costs depend on the characteristics of each individual installation.

On smaller installations in which the grain loading (the old
federal and state limits were generally 0.08 gr/dscf of stack gas) to
the collector is relatively low and the emission regulations are not
as stringent, two mechanical collectors in series, one after the
other, may be sufficient to meet the standard. The first collector
removes larger particles and the second collector is equipped with
high-energy vanes to remove finer material. The overall collection
efficiency of such a system can be as high as 97 to 98%.

Systems with particulate loading too high to be collected with
series mechanical collectors, but low enough not to require full
airflow to be sent through a scrubber, can sometimes meet the
standard with a selective collector–scrubber system. The full flow
goes through a mechanical collector where the dirtiest portion (10
to 20% of the total) is separated out and fed to a wet scrubber.
This reduces the size and cost of the wet scrubber to allow emission certification at the most economical cost.

Two different selective-flow scrubber systems—shave-off and
fractionating—are currently on the market. In the shave-off system, the multiclone outlet is designed with two outlet tubes instead of one. The inner tube allows the core of cleanest air to pass
through. The outer tube shaves off the perimeter of the outlet gases
in which, due to centrifugal action, the majority of the particulates
resides.

The fractionating system approaches the problem by removing the particulates before the airstream reaches the outlet tube. A
manifold is attached to the cyclone below the tube sheet, allowing
a controlled extraction of air from the large cyclone cavity. This
extraction aids particle disengagement from the air when it
changes direction at the exit, and the bleed-off air tends to sweep
the dust from the vessel bottom.

Most large-scale operations require a mechanical collector followed by a wet scrubber. Due to the tighter emission standards on
large boilers, all the flow is sent through the scrubber. This combination should be able to meet even the most stringent emission
standard.

The purpose of the selection process is to select the most economical control system that will still meet the regulations. In a
turnkey operation in which a design engineering firm handles the

installation, compliance with emission standards is guaranteed by the vendor. The design firm and the boiler manufacturer meet, and by considering important criteria such as fuel type, boiler size, and emission control labels, arrive at an appropriate design.

Companies that buy boilers without the aid of a consulting engineer generally fall into two categories: (1) the small firm that buys a turnkey installation from a boiler contractor, and (2) the very large firm in which all engineering is handled in-house.

In both cases, economics is the preeminent consideration in the selection of an emission control system. Buy only as much control as required. Know what fuel will be used, as emission control systems are fuel specific. Knowledge of fuel properties and proper boiler operating procedures will assure a boiler and emission control system design that meets the regulations as economically as possible.

Estimated cost information for several different size systems is presented in Table 7-4 and Figures 7-10 [5] and 7-11 [6]. This price includes the control equipment, connecting ducts, support steel, stacks, controls, settling pond (where applicable), and foundations. In general, emission control requirements are determined during the boiler design phase, and the cost of the necessary devices is included in the boiler package cost.

The operating cost for multicyclones consists of the fan horsepower required to overcome the pressure loss. The costs were calculated assuming boilers operating at 100% excess air for 6,000 hours per year with electricity at 10.5¢/kwh. This excess air is conservative, and allows for air leakage downstream of the furnace (for a new, well-operated, wet-wood-fired boiler, furnace excess air would be expected to be in the 50% range). For a wet scrubber, the operating cost consists of fan horsepower, pump horsepower,

**Table 7-4.** Pollution control systems (complete): installed costs [2]

| Boiler size (lb/hr) | System | Fan size (hp) | Installed cost (dollars) | Estimated yearly operating cost (dollars) |
|---|---|---|---|---|
| 3,450 (100 bhp) | Multicyclone | 5 | 35,000 | 1400 |
| 17,250 (500 bhp) | Series multicyclone | 30 | 72,000 | 14,000 |
| 50,000 (1,450) | Multiclone/wet scrubber | 150 | 320,000 | 150,000 |
| 100,000 (2,900 bhp) | Multiclone/wet scrubber | 300 | 470,000 | 290,000 |

Note: Costs inflated 1984 to 2008 via CPI factor of 2.1.

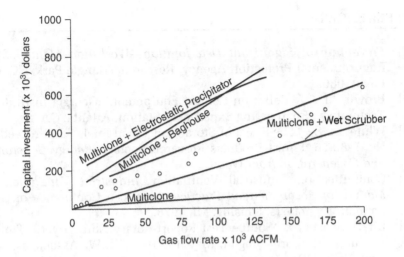

**Figure 7-10.** Capital investment for air pollution control.

and water used. The pressure loss assumed was 17″ of water column. A 5% evaporative loss was used to arrive at the water cost. The pump horsepower was assumed to be 5% of the fan horsepower. Excess air, yearly operating time, and electricity costs were the same as used with multiclones.

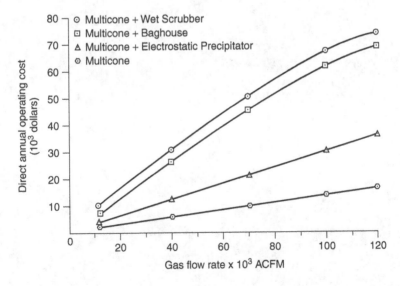

**Figure 7-11.** Operating costs of air pollution control systems.

## REFERENCES

[1]  *Prevention of Significant Deterioration—Workshop Manual,* U.S. Environmental Protection Agency, Research Triangle Park, NC, October 1980.

[2]  Brown, M. L., "Pollution Control Equipment," Georgia Institute of Technology, Engineering Experiment Station, Atlanta, GA.

[3]  White, M. S., "The Effect of Storage on Wood Fuels," in Proceeding No. P-80-26, Forest Products Research Society, *Energy Generation and Cogeneration from Wood.*

[4]  Committee on Industrial Ventilation, *Industrial Ventilation—A Manual of Recommended Practice,* American Conference of Government Hygienists, Lansing, MI, 1978.

[5]  B.W. Associates, "Source Test Report-Rotary Chip Dryer," Timber Products, Medford Oregon, May 15, 1974," B.W. Associates, Klamath Falls, OR.

[6]  Cassens, D. L., and Choong, E. T., "Fuel Values of Southern Hartwood Mill Residues," *Southern Lumbermen,* November 1976.

# CHAPTER 8

# *ENVIRONMENT AND SAFETY: RULES, REGULATIONS, AND SAFE PRACTICE*

## ENVIRONMENTAL IMPACT

A number of studies have been performed regarding the environmental impact of the growth and procurement of wood. These studies have considered harvesting operations both for timber as well as wood for energy. While the two operations are very similar, those that recover wood for energy (short-rotation tree farming) recover a greater total portion of the tree biomass and have a shorter tree life. Short-rotation tree farms may reduce animal habitat and forage, increase soil temperature, water runoff and erosion, soil compaction and aeration, and nutrient depletion. However, short-rotation tree farms have a lower impact on these factors than do normal agricultural crops.

With proper management, the harvesting of wood for energy need not impact environmental factors in any greater degree than harvesting of other timber products. Markets for energy wood could, in fact, help improve the environment by providing an incentive for conversion of unproductive forest stands and an opportunity for improved cuts in managed stands.

## PERTINENT ENVIRONMENTAL REGULATIONS

The storage and handling of wood fuel is not normally environmentally regulated, with the exception of water runoff from outdoor storage piles discharged into running streams. But the combustion products, both airborne and solid waste, are of concern to the Environmental Protection Agency. There are four major documents that relate to these areas: (1) National Ambient Air Quality Standards as amended through 1971 (NAAQS), (2) Prevention of Significant Deterioration–Clean Air Act as amended through 1977 (PSD), (3) New Source Performance Standards (NSPS), (4) and Solid Waste Management Act as amended through 1973.

The National Ambient Air Quality Standards were developed in 1971 for six pollutants that reflect thresholds of atmospheric concentrations above which the pollutants are thought to have significant deleterious effects on human health and/or on plant and animal life and property. These pollutants are particulate matter, sulfur oxides, nitrogen oxides, carbon monoxide, photochemical oxidants, and hydrocarbons. The NAAQ Standards are expressed in the terms of primary and secondary standards. The primary standards are specified to protect public health, whereas more stringent secondary standards are set to protect against effects on soil, water, vegetation, materials, animals, weather, visibility, and personal comfort and well being. The primary standards are to be met in a reasonable time as determined by the Environmental Protection Agency. The NAAQ Standards are listed in Table 8-1.

A more recent regulatory development that has greater potential to directly affect wood-burning devices is the Prevention of Significant Deterioration standard. These standards state how much of an increase a single source can add to the ambient air quality of a particular region for particulate matter and sulfur dioxide. There are three classes of PSD standards. All areas of the country will be designated as either Class 1, Class 2, or Class 3 regions for application of PSD standards. Class 1 standards allow the smallest incremental decrease in ambient air quality. Congress has already designated specific regions as Class 1. These are areas of pristine air quality such as certain size national and international parks, and national wilderness areas. Areas designated for application of Class 2 standards are allowed a larger incremental decrease in ambient air quality, although not as large as allowed in the Class 3 areas. NAAQS will act as an overriding ceiling to any otherwise allowable increment (this includes all specified NAAQS pollutants and not just particulate matter and sulfur dioxide). The PSD regulations state that all areas not classified as Class 1 will be designated Class 2. A reclassification process is involved in changing any area to Class 3. This reclassification involves,

**Table 8-1.** Ambient air standards[a]

| | Averaging period[c] | Air quality standards[b] | |
|---|---|---|---|
| | | Primary[a] | Secondary[a] |
| $SO_x$ | AAM | 80 (0.03). | |
| | 24 | 365 (0.14), 1× | |
| | 8 | | |
| | 3 | | 1300(0.49), 1× |
| Particulate | AGM | 75 | 60 |
| | 24 | 260, 1× | 150, 1× |
| | 8 | | |
| Carbon monoxide | 8 | 10,000 (9), 1× | 10,000 (9), 1× |
| | 1 | 40,000 (35), 1× | 40,000 (35), 1× |
| Oxidant (as $O_3$) | 8 | | |
| | 1 | 160 (0.08), 1× | 160 (0.08), 1× |
| $NO_x$ (as $NO_2$) | AAM | 100 (0.05) | 100 (0.05) |
| | 24 | | |
| | 8 | | |
| | 1 | | |
| Nonmethane (HC) | 8 | | |
| Hydrocarbons (HC) (as $CH_4$) | | 160 (0.24), 1× | 160 (0.24), 1× |

[a]Format for each entry is as follows. Standard $\mu g/m^2$ @ 760 mm Hg and 20°C (equivalent value, ppm). The maximum allowable exceedance rate, if any, follows. This refers to the maximum number of times per year that the standard may be exceeded. For example, 1× means that the standards may be exceeded only once per year.

[a]"National Primary and Secondary Ambient Air Quality Standards," *Federal Register* 36, #84, pp. 8186–8201.

[c]The averaging period is given in hours unless otherwise specified. AAM means Annual Arithmetic Mean Value and AGM means Actual Geometric Mean Value.

Source: Fennelly, P. F. et al., Environmental Assessment Perspectives, GCA Corp. for U.S. EPA, Washington, D.C., March 1976, p. 213.

among other items, specific approval from the governor of the affected state after consultation with the state legislature and with local governments representing a majority of the residents in the area which is to be redesignated. Table 8-2 lists the PSD standards.

Major pollution emitting facilities are required to have a PSD permit before construction can begin. The permit acknowledges that all PSD emission requirements will be met by the facility; that the proper procedures have been undertaken, such as hearings, reviews, analysis of air quality impact, and so on, that the best available control technology for each pollutant is obtained; that proper air quality monitoring will be done; and so on. Also, the PSD permit requirements include other facilities that are not specified but which have the potential to emit 250 tons or more per year of any

**Table 8-2.** Prevention of significant deterioration standards

| Pollutant | Maximum allowable increase* | | |
|---|---|---|---|
| | Class I | Class II | Class III |
| Particulate Matter: | | | |
| Annual geometric mean | 5 | 19 | 37 |
| 24-hour maximum | 10 | 37 | 75 |
| Sulfur Dioxide: | | | |
| Annual arithmetic mean | 25 | 20 | 40 |
| 24-hour maximum | 5 | 91 | 182 |
| 3-hour maximum | 25 | 512 | 700 |

*$\mu g/m^3$.
Source: *Federal Register,* Vol. 42, No. 212, 11/3/77, p. 57459.

pollutant. These permit requirements also apply to major modifications of existing facilities that cause emissions to increase by 100 tons per year or 250 tons per year depending on whether or not it is a specified or unspecified emissions source. Because of court rulings, potential emissions from a given source will be calculated after accounting for a reasonably anticipated effect of air pollution controls that are part of the facility design. Calculation of potential to emit will assume around-the-clock operation at maximum rated capacity unless the applicant can show it is impossible for the source to operate in that manner. Thus, some new sources will no longer be required to obtain PSD permits if the emissions can be controlled to less than 100 tons per year if they are on the specified emission source list or 250 tons per year for the remaining sources.

Existing emission sources that are going to be modified with the result of new emission rates over the 100/250 tons per year limits will also require PSD permits. There are some exceptions to the above requirements. These exceptions deal with emission rates that are negligible and thus fall outside the strict requirements of the law. There are two tests for this. First, if mass emissions will fall below a certain rate, the source will not need a PSD permit but must notify EPA of the proposed construction. Second, if the mass emissions are above the specified rate but the ambient air quality impact of these emissions is below the significant impact guideline, the emissions do not need to be reviewed for air quality impact.

An analysis of the minimum size boiler that would fall under PSD is given in Table 8-3. Another set of pertinent regulations are the New Source Performance Standards (NSPS). These are basically standards that limit the amount of pollutants that can be emit-

**Table 8-3.** Minimum size wood-energy boiler that would produce a level of emissions requiring a PSD analysis

| | | Size of boiler required to produce: | |
| Type of emission | Emission factors (lb/$10^6$ Btu) | 100 ton/yr of the given pollutant (lbs of steam/hr) | 250 ton/yr of the given pollutant (lbs of steam/hr) |
|---|---|---|---|
| Particulates | | | |
| Bark: | | | |
| With fly ash reinjection | 8.72 | 1,911 | 4,778 |
| Without fly ash reinjection | 5.81 | 2,869 | 7,172 |
| Wood/Bark: | | | |
| With fly ash reinjection | 5.23 | 3,186 | 7,967 |
| Without fly ash reinjection | 3.49 | 4,775 | 11,939 |
| Wood | 0.58–1.74 | 28,736–9,579 | 71,839–23,946 |
| Sulfur oxides | 0.17 | 98,039 | 245,098 |
| Carbon monoxide | 0.23–6.98 | 72,464–2,388 | 181,159–5,970 |
| Hydrocarbons | 0.23–8.14 | 72,464–2,048 | 181,159–5,119 |
| Nitrogen oxides | 0.25 | 66,667 | 166,667 |

Assumptions:   Boiler operates 350 days/year
1000 Btu/lb steam
70% boiler combustion efficiency
50% moisture content
$8.6 \times 10^6$ Btu/ton of green wood

ted per million Btu of fuel burned. The present standards regulate emissions from coal-fired utility boilers that have an energy input greater than 250 million Btu/hr and produce electricity, 25 megawatts or more of which goes to the grid for commercial use. These regulations limit particulates, sulfur oxides (SO, measured as $SO_2$), and nitrogen oxides (NO, measured as $NO_2$) as follows.

*Particulates.* May not exceed 0.03 lb/million Btu of coal burned.
*Sulfur oxides.* May not exceed 1.2 lb/million Btu of high-sulfur coal burned; a minimum of 90% of SO must be removed even though this places the emission well below the limit. May not exceed 0.6 lb/million Btu of low-sulfur coal; a minimum of 70% of the SO must be removed, even though this places the emission well below the limit.
*Nitrogen oxides.* May not exceed 0.6 lb/million Btu of high-sulfur coal burned and may not exceed 0.5 lb/million Btu for any other type of coal burned.

Only when wood is burned in combination with coal do these NSPS regulations apply to wood combustion; specifically, if a utility boiler (a) produces 25 megawatts or more of power for commercial consumption, (b) burns a mixture of coal and wood, of which the coal cannot exceed 25% of the total energy combustion in a 90-day period, and (c) has coal input exceeding 250 million Btu per hour, then the NSPS regulations for coal apply, with the modifications that the SO limit is 1.2 lb/million Btu for all coal types and there is no percentage reduction requirement.

In a practical sense, there are few installations for which NSPS regulations will apply. Standards for utility and industrial boilers that are 100% wood fired are forthcoming but not yet published. It is believed that these standards will likely focus on particulate emission. The magnitudes of particulate emissions are dependent upon the chemical and moisture content of the wood, the degree of fly ash reinjection employed in the boiler, and the boiler design and operating conditions. Control technology in the form of cyclones, wet and dry scrubbers, baghouses, and precipitators can collect up to 99.9% of the particulate matter emitted. Became of the low sulfur content of the wood, sulfur oxide emissions are minor. Nitrogen oxide emissions are also minor in a well-operated facility. Therefore, if regulations are issued on $SO_2$ and $NO_x$, they will most likely be minor and not of major concern.

The production of carbon monoxide and aromatic hydrocarbons rises sharply with poorly designed equipment and/or poor operating conditions. Therefore, attention is being given to the issue of including among the regulated pollutants carbon monoxide and hydrocarbons, particularly polycyclate aromatic hydrocarbons (PAHs). PAHs are of concern because some are known carcinogens.

Two other factors are worth noting about emissions limits. The first is location, with facilities in nonattainment areas being subject to greater regulatory hurdles. The second is dispersion modeling, which may be required for the emission source. When it comes into play, short stub stacks may be inappropriate, in particular when they are near taller buildings or equipment.

## MACT

The most recent environmental regulation affecting biomass boilers is the Boiler MACT (maximum achievable control technology), available at www.epa.gov/ttn. It covers a wide range of industrial boilers and process heaters, including those in NAICS 322, pulp and paper. The rule includes limits on PM, metals, $NO_x$, $SO_2$, and organics in stack emissions.

## Solid Waste

Ash waste from wood combustion is considered less detrimental than solid residues of coal combustion, with the only toxic substance that has been identified in wood ash being lead, present in minor quantities. Ash from wood burning is, however, considered solid waste and, as such, is subject to the provision of a Solid Waste Management Act which is administered by the State regulators. They provide the requirements for permits for solid waste handling and operation of disposal sites.

Many biomass-fired plants sell or give away the wood ash, which is used by farmers as a soil amendment. This takes it out of the "waste" category and puts it in the "product" category.

The compliance effort of the program determines waste streams and waste inventories, and monitors operation of facilities, with emphasis being given to resource recovery, reclamation, recycling, and use of residuals to eliminate the residual waste. Industrial facilities that are designed to operate their own disposal sites make application to the division for permit. Upon receipt of the application, the environmental acceptability of the proposed site is determined. Then, upon completion of the assessment and review and approval of the design and operation plan, a permit is issued for the operation of the site. Requirements for disposal of residual ash waste are provided in the rules and regulations of the act. These rules and regulations utilize criteria equivalent to that contained in subtitle D of the Federal Resource Conservation and Recovery Act of 1976. Briefly, the criteria include the following:

1. Floodplains. Facilities or practices in the floodplains shall not restrict the flow of the base flood (100 year floodplain), reduce temporary water storage capacity of the floodplain, or result in washout of solid waste, so as to pose a hazard to human life, wildlife, or land or water resources.
2. Endangered species. Facilities or practices shall not cause or contribute to the taking of any endangered or threatened species of plants, fish, or wildlife, nor result in the destruction

or adverse modification of the critical habitat of endangered or threatened species.

3. Surface water. A facility or practice shall not cause: (a) a discharge of pollutants into waters in violation of the requirements of the National Pollutant Discharge Elimination System (NPDES) permit; (b) a discharge of dredged or film material in the water in violation of Section 404 of the Clean Water Act; (c) nonpoint source pollution of waters violating Section 208 of the Clean Water Act.

4. Groundwater. A facility or practice shall not contaminate an underground drinking water source beyond the site boundary.

Runoff from open storage areas may occasionally be deemed a source of contamination to running streams and, to a lesser extent, groundwater. The absence of a sloped concrete pad beneath the pile allows a depression to form under the pile. In this area, acidic water collects (pH of approximately 5) containing lignins and other water-soluble components of the stored wood fuel. After a heavy storm, the collected water (bearing high contaminant concentrations) is flushed from beneath the pile. This solution, because of its high acidity, may be damaging both to groundwater aquifers and aquatic environments. Provision of a sloped concrete pad beneath open storage areas will minimize problems of contaminated runoff.

## SAFETY

The two major divisions of safety are safety of personnel that are involved with the fuel supply and the safety of the fuel supply itself. Personnel safety will come mainly under the jurisdiction of OSHA (Occupational Safety and Health Administration), which is a regulatory agency with fine-assessing capability. Fuel supply will be under the jurisdiction of codes such as those promulgated by the National Fire Protection Association and the National Electrical Codes, which of themselves are not regulatory agencies.

### Personnel Safety

OSHA was created by the Occupational Safety and Health Act (Public Law 91-596). This law is considered landmark legislation because it is the first national safety law in the history of the nation. It establishes standards that require each employer to provide his workers with workplaces free of recognized hazards that could cause death or serious injury. It also requires the employee to comply with all safety and health standards that apply to his job. Most of the rules and regulations that apply to personnel involved with a wood fuel supply are contained in "General Industry Safety and Health Regulations" Part 1910.

The applicable rules and regulations will depend upon each specific site, how extensively the wood fuel is processed, and the extent of the handling necessary to convey the fuel to the furnace. However, there are eight basic categories of injuries. These are: (1) electric shock or electrocution; (2) burns; (3) overexertion; (4) poisoning by inhalation, ingestion, absorption, and so on; (5) being caught in; (6) being struck by; (7) striking against; (8) and falls, both from the same level and from a different elevation. The objective of applicable rules and regulations is to provide equipment and processes that eliminate the potential for any of these eight types of accidents. Conveyor safety is an issue with solid fuels, as well as care in confined-space entry, and potential for engulfment by solid fuel in bins and piles.

As a first step, all mechanical power transmission devices must be guarded; gears, chains, sprockets, belt drives, and so on, should be enclosed. As a test for adequacy, wherever the power transmission drive may be located, if you can reach over, under, around, or through the enclosure and contact moving parts, then the guarding device is not adequate. Make sure that the machine is off before you try this!

The same criteria can be applied in determining whether the equipment's point of operation hazard is effectively guarded. If a hand or finger, through any route, can reach an area where work is performed on a material, then the potential for an accident exists and the machine is not adequately guarded. OSHA 2057 discusses the principles and techniques of mechanical guarding.

All equipment both stationary and mobile must meet certain standards. For example, the cab area of a front-end loader must be protected by mesh screens that prevent entry of obstacles. Also, crane and loader stability must be in accordance with the American National Standards Institute's (ANSI) safety code for cranes, derricks, and hoists. Chippers should not be opened until a drum or disk is at a complete stop and this should be accomplished by interlocking. Machines for hogging wood should be so designed that at no position on the rim of the chute should the distance of the blades be less than 40 inches. A safety harness and lifeline must be worn by any workman at or near the spout unless the feed chute on the hammermill is guarded.

Personal protective equipment must be provided for employees. These include face protection, safety helmets, safety shoes, gloves, dust masks, and hearing protection if the conditions warrant. Also, operator instruction programs for the safe operating limits of the equipment are required.

Depending upon the size of the operation, storage and conveying systems can be either elaborate or simple. In all cases, though, there will be regulations governing work in and around storage bin areas and conveying equipment, and guarding of their ladders, walkways, and moving and rotating equipment. In addition to OSHA requirements, the American National Standards Institute directly addresses material-handling systems in its "Standards for Conveyors and Related Equipment." This document deals with guarding moving parts, types of guards, control functions, platforms, operation, and maintenance procedures.

An aspect of safety that is addressed both by OSHA and insurance carriers is that of operator education and maintenance-personnel instruction. Once equipment has been engineered to comply with safety standards and has been set up correctly, nearly every accident that occurs thereafter can be traced to the unpredictable element of the system—the people. It requires constant continuous management effort to ensure that the operators adhere to operating procedures and that maintenance personnel do not take shortcuts in their work.

In summarizing personnel safety, it is necessary to analyze the equipment, the system, and the plants for the potential of any of the eight different accident types occurring. Where the potential is found, the hazard must be eliminated with designs that use OSHA standards as guidelines and sources of technical information. Safety education programs must be provided on a continuing basis.

## Fuel Supply Safety

The major threat to wood fuel is fire. Fires can originate from both external sources and from spontaneous combustion. Fires from external sources (such as smoking, electrical shorts, etc.) can occur in any size fuel wood. Internally initiated fires from spontaneous combustion can occur in bark piles and piles of very small chips and/or sawdust (fines) since the spontaneous combustion generated heat is a function of the surface area/mass ratio and packing density. Whenever highly resinous dry fines are mixed with wet chips, the potential for spontaneous ignition occurs. There are instances of chip piles recording temperatures near the boiling point of water and subsequently experiencing spontaneous combustion.

Recommended safe practices for storage and handling of wood chips and logs have been published in the National Fire Codes of the National Fire Protection Association. Experience has shown that the principal factors that allow lumberyard fires to reach serious proportions are large undivided piles, congested storage conditions, delayed fire protection, inadequate fire protection, and ineffective firefighting tactics.

A positive fire prevention program is required and should include selection, design, and arrangement of storage yard area and handling equipment based upon sound fire prevention and protection principles, facilities for early fire detection and alarm transmission, fire extinguishers, fire monitor nozzles, fire lanes to separate large piles and provide access for effective firefighting operations, separation of yard storage from mill operations, regular yard inspection by trained personnel, including an effective fire protection maintenance program, and a fire hydrant system connected to an ample water supply. Also, pile height should be limited, as heights in excess of 20′ seriously restrict effective extinguishing operations and increase the chance of spontaneous combustion. Of course, smoking, cutting, welding, open fires, as well as unauthorized persons, should be prohibited and all electrical equipment and installations should conform to the National Electrical Code or National Electrical Safety Code.

Storage precautions that may be taken to guard against spontaneous combustion fires in chipped or hogged wood piles are:

Keep all refuse and old chips out of the pile base
Keep whole-tree chips free from bark chips
Use whole-tree chips first
Choose a clean storage site
Keep buildup and reclaiming of a wood pile to a maximum
    turnover time of 1 year

Limit pile size

Install thermocouples during pile buildup or provide other means
    for measuring temperatures within the pile

Avoid concentration of fines

Wet the pile in dry weather

Minimize vehicle compaction in the chip pile

Provide fire monitor nozzles

Provide covered storage areas with automatic sprinkling protec-
    tion

There are safety issues in biomass storage in silos and fuel
storage bins. Confined-space entry procedures, including verify-
ing viability of air in the vessel and providing ventilation (*note
that fresh silage gives off nitrogen oxides that can be fatal*) must be
used for personnel working inside them. If ratholing and/or bridging occurs in a storage vessel, it is best to fix the problem via vi-
brators rather than sending personnel into areas where they are
subject to engulfment. If severe bridging and compaction occurs, it
is best to call in outside specialists to get the silo back in operation.

Another aspect of wood fuel safety relates to combustion.
Conventional fuels such as coal, natural gas, and oil have a relatively constant heating value. Consequently, it is practical to establish a desired fuel feed rate and to supply the proper amount of
air for stoichiometric combustion plus an appropriate level of excess air. Wood presents a more difficult problem in that its Btu
content for a given volume or weight of fuel can vary widely. The
species of the wood and percent of the wood bark in the chips will
affect the heat content; however, the major factor is the amount of
moisture. Establishing the correct air-to-wood fuel ratio is performed manually by the boiler operator by observing the flame
and monitoring combustion temperature. Wood fuel feeding
equivalent is based on volume. For example, an auger feed unit
will deliver a certain volume of material at a given rotational
speed of the auger. The boiler operator soon learns the setting of
combustion air flow for a given auger speed. This technique works
reasonably as long as the wood fuel is relatively uniform in its
characteristics. A safety problem can be created if the fuel characteristics change suddenly.

For example, if a system is running on green wood and dry
fuel is introduced, the volume of material will remain constant

but the heat content can nearly double. To combust this dry material correctly, the volume of fuel should be reduced while keeping the combustion and air volume the same. In practice, a boiler operator may not realize that they fed a slug of dry material. Since there is not sufficient air to completely burn the wood, the result is that the combustion chamber, the heat transfer section, and even the breeching, chimney, or boiler, may be filled with a combustible gas. This condition is analogous to having a leaking gas valve that fills the boiler with natural gas. A boiler filled with wood gas presents a similar potentially explosive situation. When wood heating value goes up significantly, the stack will go black from soot, and combustion temperatures will shoot up.

There is presently no remedy for this condition except operator training and vigilance, and setting high limit alarms on stack thermocouples. An oxygen continuous emission monitor or CO monitor could also be used with alarms. Certainly, new boilers should be observed carefully during the startup phases and any time a new supply of fuel is received. While major explosions are quite rare, minor ones do occur and can result in personal injury and equipment damage.

Other items of safety concern are fires in emission control equipment due to the accumulation of unburned carbon; spark carryover where burning wood particles are carried through the boiler and out the furnace stack; ash and flyash that contain hot, glowing particles that are dumped on combustible materials or smolder or ignite on contact with air; dust explosions with very dry sander dust or wood shavings, and housekeeping fires that occur from wood dust accumulation.

# CHAPTER 9

# BIOMASS FUEL SUPPLY AND PURCHASING

## DETERMINING FUEL SUPPLY

Procuring biomass is done via brokers or directly via the operating firm's personnel. Consulting foresters frequently do the legwork for wood supply. For agricultural waste, the local agriculture extension agent may be a good resource, or one can directly contact firms that process agricultural feedstocks, such as nut shelling operations and other food processing operations, as they represent a concentrated source, similar to sawmills.

Agricultural waste cost will vary by feedstock, based on location and whether any other end use currently exists, and on cost of landfilling if that is the current disposal method.

The cost of wood fuels varies from inexpensive mill residue (chips, sawdust, bark, and shavings) to dry wood pellets. Price is directly related to the type of wood (softwood, hardwood), moisture content, hauling distance, local supply and demand situation, and quantity purchased.

This section explains how a survey was taken to determine the mill residue supply in Georgia. The methodology is applicable to any locale. There are references that give a methodology for determining the mill residue produced given a quantity of production, but it is largely dependent on doing a study on the mill in question. As more mill residue is utilized and sold, mill operators become aware of how much residue their mill produces and how much it is worth. In short, word gets around, and prices go up.

Mill residue is the cheapest fuel wood available. The variability of the material (sawdust, chips, boards, blocks, bark, and slabs) and high moisture content (50%) may make handling, storage, and

*Biomass and Alternate Fuel Systems*. Edited by McGowan, Brown, Bulpitt, Walsh
Copyright © 2009 American Institute of Chemical Engineers, Inc.

burning of this fuel costly. But its low price ($15–20 per ton) can make it an attractive fuel. One study said only 8% is unused. However, this varies with location and with the state of the economy and level of new construction. The majority of it is sawmill residue and is burned as boiler fuel or used in making fiber products.

As a first step in establishing a database for wood fuel supply, it is necessary to evaluate the many sources currently available. Generally, there is much information already available that can be used to good advantage.

The first source to check is the State Forestry Commission and the USDA Forestry office. Part of their responsibility is maintaining wood supply data. Most of the data that they publish concerns the amount of growing timber and removals on a county-by-county basis. This can give a good estimate with regard to the amount of mill residue or whole-tree chips produced within a certain area. This does not necessarily mean that the number indicated is what is available for sale as fuel. It has been found that many of the operations are beginning to use more and more of the waste product for the purposes of steam production and heating or drying their own products. Still, many are giving much of their residue away or carting it to the dump. Of course, producer utilization depends on the market conditions existing in the area.

A good place to look for suppliers of mill residue is in the State's manufacturing directory. This directory is probably located in the local library, or is obtainable from the State Department of Industry and Trade (or Economic Development). It is much better to look at a current edition because of changes in ownership and fluctuations of the forest product industry's business with the economy. The companies in the directory are listed several ways, but the best for this purpose is the listing by NAICS (North American Industry Classification System). This replaced the older SIC codes. The NAICS codes related to wood fuel producers are:

| NAICS Code Number | Classification |
|---|---|
| 321113 | Sawmills |
| 321912 | Hardwood dimension and flooring |
| 32191 | Millwork |
| 321920 | Wood pallets and skids |

There are other sources of residue, such as pulp and paper plants, but, generally, they use all that is produced and buy more. The four codes listed above are the ones that we have addressed and from which a good reply was received.

Supply planning should be limited to areas that may be capable of supplying the quantity necessary for an industrial or commercial operation. There are three basic types of wood fuel for industrial fuel supply survey purposes: whole-tree chips, wet or dry mill residue, and wood pellets.

The whole green tree chips are produced by running the whole tree through a large, expensive chipper.

While more expensive than mill waste, whole-tree chips represent a significant expansion of the supply, and can be considered a good source of fuel.

Wood mill residue can be found at many sawmills, furniture manufacturers, pallet and plywood manufacturers, and other forest product industries. Mill residue is a difficult entity to quantify, especially if the mill operator is not currently selling waste. The operators know they have a huge pile of it every day, or it is stacked up out back, but usually have only a vague idea of how much there is unless it is being sold on a contract basis.

Wood pellets are being sold for export and use in residential stoves. They are too expensive to compete with gas and oil for industrial applications on a price basis.

## PURCHASING AND CONTRACTING FOR WOOD FUELS

After conversing with mill residue and whole green tree chip suppliers, it was found that contracts and specifications ran the full spectrum. Most everyone polled had some sort of verbal agreement that had been in force for some time. Others had written contracts that were 3 to 4 pages long all the way up to 35 pages.

Most of the written contracts in the mill residue area were called waste purchase agreements. They ran from 1 to 5 years maximum. As contracts, they were loosely written. For example, they were reviewed semiannually and could be changed within 30 to 60 days if both parties agreed; otherwise, the contract was void. These contracts generally contained FOB price at the mill and a freight rate allowance. Both of these were subject to change at any time with the freight rate indexed to the prevailing freight rate.

Note that the cancellation clause in a contract is critical for the seller, as a downturn in construction and use of lumber can require the plant to cut back or shut down on relatively short notice. Similarly, the wood buyer may also have plant production cutbacks, and need the ability to give notice of reduced purchase levels.

The whole-tree chippers worked both from a verbal (purchase order agreement) and a written contract. The verbal contract with the big mills was most prevalent. The whole-tree chippers were in the 100-ton/day range of production and could more easily guarantee the supply to meet the demand without the need for the written contract. Note that whole-tree chippers are subject to weather constraints, which vary with locality. For example, extended heavy rain would rule out production for a few days in the south, and "mud season" in the Northeast would also slow or stop production of whole-tree chips.

The wood brokers that were contacted also used verbal and written agreements. One wood broker said that there was no standard contract. He said every one was different. The one common factor was that they were long—about 30 or more typewritten pages.

Another wood broker was not using a written contract at present, but was considering going to one. He also felt that each contract would have parts to it that would make it unique.

Overall, it appeared as though two important underlying factors were having a large impact on the use of written contracts. The two factors, supply and cost, are a result of the volatility of the wood market.

# CHAPTER 10

# FUEL-SWITCHING FEASIBILITY STUDY METHODOLOGY

An engineer or manager performing a feasibility study for biomass energy is often unaware of the many aspects to be considered. The same person may be equally uncertain as to where to begin the study and in what order each aspect should be investigated. The purpose of this chapter is to present a methodology that organizes the data collection process in a form that has been successfully used in performing many biomass and fuel-switching feasibility studies.

The step-by-step guide presented here will be useful for those totally unfamiliar with biomass energy as well as for those persons who have already been involved in studies. The flow of the data collection steps is set up so as to improve the efficiency of the study by limiting the collection of data until certain go/no-go indications have been passed. Effort expended in analyzing copious amounts of data is thereby prevented. The format is useful for setting up or refurbishing any fuel system, be it biomass, coal, or other fuel type.

This chapter covers only the broad steps required to complete a study and their respective order of processing. Details relating to individual systems and biomass fuel characteristics can be found in other chapters of this handbook.

The flow chart shown in Figure 10-1 depicts the order of the steps to be completed. The starting point must be a detailed assessment of the energy requirements. This assessment will be used to determine system characteristics, fuel type, fuel supply require-

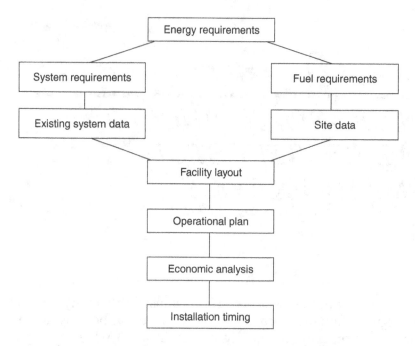

**Figure 10-1.** Order of investigation for feasibility studies.

ments, and so on. Therefore, it must be detailed in both short- and long-term aspects.

## ENERGY REQUIREMENTS

Energy requirements can be divided into three general categories:

1. Load Cycle
2. Hours of Operation
3. Annual Consumption

Each category will be used for different aspects of the study. Abnormalities in energy demand should be equally detailed.

Peculiar to each individual operation, the load cycle has no defined time base. It can be as short as a few hours or as long as several months. Consider the textile drying operation shown in Figure 10-2. For simplicity, energy demand is shown as a percent of available capacity. In detailing a particular load cycle, however,

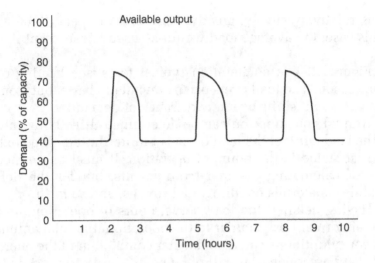

**Figure 10-2.** Industrial process steam with peak demands.

a more quantitative form (such as pounds per hour of steam) should be used. For this example, a fairly constant base load is experienced with a periodic peak load, for example, due to opening and closing of a steam valve for a dryer.

Figure 10-3 depicts a demand cycle having a long time base. A common situation for industrial processes having some peak modulation of a nonperiodic form, the overall average load re-

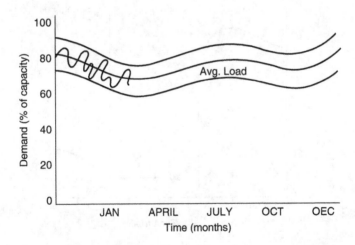

**Figure 10-3.** Industrial process steam with steady demand.

mains relatively steady, greatly exceeding the peak fluctuations. In this case, the average load would be used to represent the load cycle.

Figure 10-4 is an iteration typical for space heating over a given season. The load cycle occurs on a daily basis, but from season to season it shifts over a wide range of demand.

Annual consumption can be determined directly by integrating the load curve if the load cycle is uninterrupted. The next data to be tabulated is the hours of operation. It must be defined not only for calculating the fuel requirements, but for the delivery schedules, materials handling, fuel storage, and so forth.

Having detailed the load curve, hours of operation, and any other abnormalities peculiar to the operation, the total annual energy consumption is calculated. After completion of the energy requirement assessment, it will often be necessary to proceed to the next step before a conclusive decision can be made.

The energy assessment gives insight into several areas. Any plant contemplating an alternate energy facility must operate a large percentage of the available time. This is true because savings to pay for the new system are generated by the use of less-expensive alternate fuel.

Using fuel by operating more hours produces more savings. A large expenditure for an alternate energy system that sits idle is a

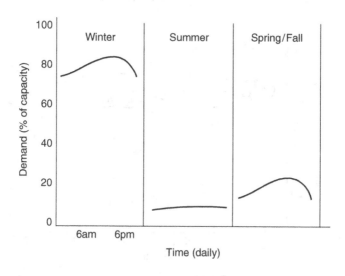

**Figure 10-4.** Space heating/hot water, seasonal demand.

waste of capital. Biomass energy systems are most practical in applications operating 7000 hours or more per year and least practical in applications operating 3000 hours or less.

For the same reasons, the size of the system is important. Large systems can produce more savings by using greater amounts of low-cost alternate fuel. There will be a lower limit on the size of biomass fuel systems below which economical justification cannot be made.

The load cycle could also influence the feasibility of the system. Wet-wood systems, due to the thermal merits of the furnace refractory, tend not to adjust to rapidly changing load patterns. Where swing loads are anticipated, the best arrangement is to use a smaller base-loaded biomass boiler with a gas/oil boiler supporting the load swings.

## SYSTEM AND FUEL REQUIREMENTS

Energy requirements can be translated into system and fuel requirements. If the investigator has little or no experience in biomass fuel characteristics, assistance will be required. Information on biomass handling, storage, supply, and combustion is presented in other chapters of this book. There are some "rules of thumb" calculations and charts to translate the energy requirements into fuel requirements and systems considerations (see Chapter 3). Generally, a boiler efficiency of 65% is used for wet-wood firing. The fuel requirement for a 500 hp boiler is (assuming green wood fuel at 4500 Btu/lb × 2000 lb/ton = 9,000,000 Btu/ton):

500 hp × 34.5 lb/hr/hp × 1000 Btu/lb × 1/.65 × 1 ton/9,000,000 Btu

= 2.95 tons/hr of green (50% moisture content wet basis) fuel

Thus, a 500 hp boiler would consume approximately 2.95 tons/hr operating at full load. This assumes that 1000 Btu/lb are required for boiling. This approximation circumvents laborious, theoretical analyses and is practical for feasibility determination.

On the system side, a determination is needed of the boiler or furnace size necessary to meet the peak and annual energy requirements. The type of unit selected must be able to respond to the load cycle. The fuel feeding requirements will also evolve from this step. The energy requirement details can be translated into annual biomass consumption, daily and weekly requirements, and storage or standby requirements.

Completion of this step will yield some solid go/no-go indicators such as:

- Suitability of a biomass system to the plant load swings
- Potential savings produced by the energy demand
- Compatibility of biomass fuel delivery with hours and type of operation
- Availability of the biomass fuel supply

## EXISTING SYSTEM AND SITE DATA

If the system and fuel requirements seem plausible, the next step is to develop the available options by examining the existing system and site in detail.

On the system side, the possibilities of boiler conversion must be explored. Moreover, for conversion to be a real option, the system must be able, or have been able, to burn solid fuel. Solid-fuel furnaces are equipped with the large furnace volume and floor grates necessary to support solids combustion. Package gas/oil boilers have neither the combustion volume nor the grates necessary to support biomass firing. If the existing boiler is amenable to conversion, remember that boilers designed for coal firing will experience some degree of derating when converted to biomass fuel (due to the larger volume of off-gases generated by combusting moist biomass fuel). Generally, the boiler percentage derating will range from 25 to 30%. If the ability to convert is lacking, an entirely new piece of equipment must be considered.

Confirmation of the fuel supply requires contact with the local agriculture and/or forest industry. Plant personnel may be familiar with sawmill operators, loggers, and/or chippers in their area that could serve as wood fuel suppliers. Additional data may be gained by contacting the State Forestry Commission (see Appendix 2) and agricultural extension agents.

For the site, items to be investigated include confirmation of the fuel supply, transportation alternatives, receiving area configuration, storage alternatives, and overall available space. When contacting potential suppliers, it may be possible to learn if and how biomass is being transported at present. Some areas have truckers with dump trucks or live-bottom vans in operation. In most areas, wood waste is handled in conventional "rag top" trailers. Agricultural waste is transported in a variety of vehicles, including metal-mesh trailers. The available method of fuel transportation will add insight into the type of receiving area necessary. If self-unloading alternatives such as live-bottom vans are available, the fuel receiving area will be greatly simplified. Otherwise, a fuel receiving area must be designed. Its specification will depend on the amount of biomass to be received. Small systems will be most economically served with a "bobcat" or equivalent front-end loader, whereas large systems will best be served by a truck dump. Storage alternatives, discussed earlier in this book, consist of selected amounts of open and covered supply. The wood storage must be consistent with the plant characteristics and space availability.

The question of space is of prime importance when considering a biomass energy system. Not only must there be sufficient space for the new boiler house or other fired equipment, but storage space and vehicle accessibility must be provided. In locations where space is limited, a silo provides the best alternative for covered storage. Open storage can then be placed at some remote location.

Any number of go/no-go indicators may surface from these investigations and there is no defined order for the data collection during this step. Most investigators tend to examine fuel supply and storage areas first, probably out of greater concern for fuel availability. From a financial viewpoint, the biomass supply and cost may provide an excellent benchmark in fuel cost savings, projecting a maximum capital outlay based on company return on investment and payback policies. This in itself is a good go/no-go indicator. Yet the examination of storage space is also important since it constitutes the largest area required by the system.

## FACILITY LAYOUT

If indications are good, it is now time to make selections from all of the options and prepare the proposed facilities layout. A complete breakdown of hardware and implementation requirements will be needed for the following steps.

The facilities layout should cover every aspect of the system from biomass supply through pollution control. With confirmation of a fuel supply, the delivery and receiving modes can be selected. This should include transportation to the site as well as unloading equipment, transport to storage, and so on.

If fuel processing is called for, the hogging and screening equipment and associated handling should be detailed. All storage areas should be located and sized to meet peak and contingency requirements, and materials handling in and out of storage should be specified.

The combustion system is based on either a conversion of existing equipment or a new installation. Ash removal should not be overlooked, and plans and equipment for this should be selected.

Pollution controls and alternate back-up fuels, if included, should also be incorporated into the proposed plan.

## OPERATIONAL REQUIREMENTS

The operational plan will go hand in hand with the facilities layout and will include personnel additions, maintenance considerations, spare parts, and back-up requirements. Unlike gas/oil systems, a biomass system will require a full-time operator.

## ECONOMIC ANALYSIS

Next, it is time to develop cost estimates of the proposed facility layout. Project economics can then be run, and can be as simple or detailed as company policy dictates. The payback and return on investment generated will most likely be the most significant go/no-go indicator, along with the total capital required for the proposal.

## INSTALLATION PLANNING

The final step that may or may not be necessary for the particular study (but which makes an excellent complement) is an installation timing investigation. Part of the facilities layout will have addressed various pollution control options, and approval of permits can require considerable lead time. The timing of permit applications, shutdown scheduling, construction time, and equipment delivery lead times will be important factors if the system is implemented.

# CHAPTER 11

# ECONOMIC ANALYSIS OF BIOMASS COMBUSTION SYSTEMS

The economics of biomass combustion systems are addressed by analysis of those factors common to any financial investment. The analysis focuses on the following topics:

- Capital Investment
- Operating Costs
- Taxes and Insurance
- Fuel Costs
- Annual Income/Savings
- Payback and Return on Investment

Budget capital costs for a proposed system are obtained by surveying vendors of biomass fuel equipment and from estimations derived from construction handbooks. Major items are budgeted with installation included. All items include a full complement of accessory items and are "turnkey" estimates ready for operation.

Operating costs are the most difficult to estimate; however, their impact on the analysis is not large. Biomass fuel systems generally do experience higher operating costs because of their lesser degree of automation and increased personnel requirements. One successful approach to estimating operating costs is derived by surveying users of various systems. From the compiled cost data, a generalized cost is then estimated on the basis of Btu output.

Property tax and insurance costs vary depending on location, rate classifications, carriers, and so on. Specific detailed information should be obtained in this area when final design is initiated. Again, the survey approach can be used for this estimation.

Fuel costs are separated from the operating costs, as they have the greatest impact on the analysis.

With the cost information assembled, the analysis can be performed on the basis of annual cash flows, taxes, and so on. If conventional fuel systems and their respective costs are being displaced, savings (rather than income) are generated. If the biomass is being hauled off at present, then savings from avoiding the cost of hauling and tipping fees must be figured in.

## TAX CONSIDERATIONS

Tax consequences have a significant impact on business investment. An effort needs to be made to approximate these effects. Investment tax credit and depreciation schedules should be calculated. Federal and state taxes should be applied to the gross income after deducting depreciation. Investment tax credits may be available for application to the ordinary tax liability. It can be assumed that losses created by tax credits or depreciation will provide tax relief for other corporation income and, therefore, can be treated as income (as opposed to carrying the loss forward to later years). It also should be assumed that the maximum tax rate applies to the project even if income is below the maximum tax levels, as other corporation income would be involved.

### Depreciation

The depreciation schedule should be based upon current IRS guidelines for business investments and expenses. In the example shown here, tangible property used is depreciated at a rate of 200% using the declining balance method. For simplicity, the entire investment capital may be considered as 20-year-life tangible property. In a more detailed analysis, shorter-life equipment such as front-end loaders, conveyors, and other minor items would be depreciated for 5- to 10-year lives under straight-line methods; however, the effects of those refinements would be minor.

## Investment Tax Credit

Investment tax credits have applied in the past and may be available for current biomass installations. It is critical that the user determine what tax credits exist, and what the rules are for their use. The example assumes this is the case, for equipment having a life greater than 7 years, with the credit equal to 10% of the total cost of the equipment. Buildings and land do not qualify for the credit. The entire capital investment (less buildings) can be considered to have a life of over 7 years for investment tax credit computations.

In the past, some biomass projects qualified as an alternate fuels project that qualified for an energy investment tax credit of 10% of the total cost. The same capital basis can be used for this computation as for the previous credit.

Limitations applying to tax credits are based on the corporation's total tax returns. It may be assumed that all credits will be allowed with no limitations. Again, a further refinement of these computations should be made prior to financial commitments. Even with some broad assumptions, the computations can become tedious for a 20-year analysis. It is advantageous to utilize a computer to generate annual cash flow tables.

The computer can speed the calculations; however, its power lies in being able to generate summaries of the analysis while varying one or more inputs. This allows the user to see the sensitivity of the project's economics to changes in estimated input(s). Confidence levels in the project are then generated regardless of estimation errors.

Biomass system analyses have great sensitivities to fuel prices, as might be expected. Sensitivity to investment capital is usually slight, and to operating costs, negligible.

The following is a brief summary of one analysis program that has been successfully utilized for biomass combustion systems. Program inputs and outputs are shown in the following outline for a wood-fired system.

I. DATA INPUTS
   A. Capital Investment
      Inputs are taken for:
      1. Total capital of project
      2. Equity (amount not financed)
      3. Amount qualifying for tax credit
      4. Amount to be depreciated

| | | |
|---|---|---|
| Capital Investment | $1,470,000 | |
| Capital Equity | $0 | |
| Tax Credit Capital | $1,338,750 | |
| Tax Credit Rate | 20. 0% | |
| Depreciable Capital | $1,470,000 | |
| Income Tax Rate | 39.0% | |
| Interest Rate | 7.0% | |
| Plant Life | 20 | |
| First Year Additional Operating Cost | $ 43,260 | |
| Additional Operating Cost Escalation Rate | 5.0% | |
| Wood Fuel Cost, delivered | 15.00 | $/ton |
| Wood Fuel Consumption (First year) | 12,100 | tons/yr |
| Natural Gas Price | 10.0 | $/MCF |
| Natural Gas Consumption (First year) | 54,850 | MCF/yr |
| Oil Price | 4.00 | $/gallon |
| Oil Consumption | 139,300 | gallons/yr |
| Operating Cost Escalation Rate | 5.0% | |
| Wood Cost Escalation Rate | 5.0% | |
| Fuel Escalation Rate, years 1 -10 | 5.0% | |
| Fuel Escalation Rate, years 11 -N | 5.0% | |

**After-Tax Discounted IRR      39.1%**

Values in table shown in thousands of dollars

| Year | 0 | 1 | 2 | 3 | 4 | 5 | 6 | 7 | 8 | 9 | 10 |
|---|---|---|---|---|---|---|---|---|---|---|---|
| Capital Payment | | ($139) | ($139) | ($139) | ($139) | ($139) | ($139) | ($139) | ($139) | ($139) | ($139) |
| Additional Capital Cost | | ($43) | ($45) | ($48) | ($50) | ($53) | ($55) | ($58) | ($61) | ($64) | ($67) |
| Wood Fuel Cost, delivered | | ($182) | ($191) | ($200) | ($210) | ($221) | ($232) | ($243) | ($255) | ($268) | ($282) |
| Total Cost | | ($364) | ($375) | ($387) | ($399) | ($412) | ($426) | ($440) | ($455) | ($471) | ($487) |
| Current Fuel Cost | | ($1,106) | ($1,161) | ($1,219) | ($1,280) | ($1,344) | ($1,411) | ($1,482) | ($1,556) | ($1,634) | ($1,715) |
| Gross Savings | | $742 | $786 | $832 | $881 | $932 | $986 | $1,042 | $1,101 | $1,163 | $1,228 |
| Depreciation (DDB 200%) | | $147 | $132 | $119 | $107 | $96 | $86 | $78 | $70 | $63 | $57 |
| Depreciation (Straight Line) | | $77 | $74 | $70 | $67 | $64 | $62 | $60 | $59 | $58 | $57 |
| Depreciation | | $147 | $132 | $119 | $107 | $96 | $86 | $78 | $70 | $63 | $57 |
| Income Tax | ($268) | $232 | $255 | $278 | $302 | $326 | $351 | $376 | $402 | $429 | $457 |
| Cash Flow | $268 | $510 | $531 | $554 | $579 | $606 | $635 | $666 | $699 | $734 | $771 |
| Discount Rate | 1 | 0.7187 | 0.5166 | 0.3713 | 0.2669 | 0.1918 | 0.1379 | 0.0991 | 0.0712 | 0.0512 | 0.0368 |
| Present Value | $268 | $367 | $274 | $206 | $155 | $116 | $88 | $66 | $50 | $38 | $28 |
| Net Present Value | $268 | $634 | $909 | $1,115 | $1,269 | $1,385 | $1,473 | $1,539 | $1,589 | $1,626 | $1,655 |

| Year | 11 | 12 | 13 | 14 | 15 | 16 | 17 | 18 | 19 | 20 |
|---|---|---|---|---|---|---|---|---|---|---|
| Capital Payment | ($139) | ($139) | ($139) | ($139) | ($139) | ($139) | ($139) | ($139) | ($139) | ($139) |
| Additional Capital Cost | ($70) | ($74) | ($78) | ($82) | ($86) | ($90) | ($94) | ($99) | ($104) | ($109) |
| Wood Fuel Cost, delivered | ($296) | ($310) | ($326) | ($342) | ($359) | ($377) | ($396) | ($416) | ($437) | ($459) |
| Total Cost | ($505) | ($523) | ($542) | ($563) | ($584) | ($606) | ($629) | ($654) | ($680) | ($707) |
| Current Fuel Cost | ($1,801) | ($1,886) | ($1,986) | ($2,085) | ($2,189) | ($2,299) | ($2,414) | ($2,534) | ($2,661) | ($2,794) |
| Gross Savings | $1,296 | $1,363 | $1,443 | $1,522 | $1,605 | $1,693 | $1,784 | $1,880 | $1,981 | $2,087 |
| Depreciation (DDB 200%) | $51 | $46 | $42 | $37 | $34 | $30 | $27 | $25 | $22 | $20 |
| Depreciation (Straight Line) | $57 | $57 | $57 | $57 | $57 | $57 | $57 | $57 | $57 | $57 |
| Depreciation | $57 | $57 | $57 | $57 | $57 | $57 | $57 | $57 | $57 | $57 |
| Income Tax | $483 | $511 | $541 | $572 | $604 | $638 | $674 | $711 | $751 | $792 |
| Cash Flow | $813 | $857 | $903 | $951 | $1,002 | $1,055 | $1,111 | $1,169 | $1,231 | $1,295 |
| Discount Rate | 0.0264 | 0.0190 | 0.0137 | 0.0098 | 0.0071 | 0.0051 | 0.0036 | 0.0026 | 0.0019 | 0.0014 |
| Present Value | $21 | $16 | $12 | $9 | $7 | $5 | $4 | $3 | $2 | $2 |
| Net Present Value | $1,676 | $1,692 | $1,705 | $1,714 | $1,721 | $1,727 | $1,731 | $1,734 | $1,736 | $1,738 |

**Figure 11-1.** Sample economic analysis.

   B. Depreciation (per IRS regulations)

One of these selections can be made for depreciation:

1. Straight-line calculation method
2. Declining balance method
3. Sum-of-the years digits method

If declining balance is asked for, an input is made for the acceleration rate (i.e., 200%).

   C. Rate Inputs

Various rates can be selected as follows:

1. Tax credit rate
2. Interest rate on borrowed capital
3. Rate for inflation (discount rate)
4. Operating cost escalation rate
5. Fuel cost escalation rate
6. Revenue escalation rate (two inputs are allowed, one for years 1–10, one for years 11–$N$)

Other Inputs

1. Life of the analysis (years)
2. Title of the project
3. Selection of whether returns and paybacks are to be calculated on a total capital or equity only basis. (This is useful if a project has some government funding.)
4. Selection of one variable for a sensitivity analysis (if a sensitivity analysis is desired)

Eligible variables include:

1. Capital
2. Interest rate
3. Life of project
4. Fuel cost (wood price)
5. Fuel consumption (wood consumption)
6. Current fuel cost (natural gas price and consumption)
7. Current fuel cost (oil prices and consumption)
8. Fuel escalation rate (wood escalation rate)
9. Current fuel escalation rate, year 1–10
10. Current fuel escalation rate, year 11–$N$

If a variable is selected, an input is made for how many entries will be made for the variable and for those entries (i.e., through inputs for wood price; $8.00, $10.00, $12.00)

II. OUTPUTS

Two main outputs are available:

1. The year-by-year life cycle cash flow table
2. The sensitivity table

The life cycle analysis is prefaced by a listing of all input parameters and selections. Each year has cash flow columns designated as:

1. Year
2. Payment on borrowed capital
3. Operating cost (additional cost for wood systems)
4. Fuel cost (wood cost, annual)
5. Total cost
6. Current annual fuel cost
7. Gross savings
8. Depreciation
9. Tax
10. Net income after taxes
11. Present value (discounted cash flow)
12. Net present value (summation of discounted cash flow)

Final outputs are:

1. Discounted cash flow return on investment
2. Simple payback
3. Discounted payback

Capital expenditure and tax credit is assigned in year "0" (assumed to be the construction period with no operational income or costs)

A sample analysis has been performed and is shown in Figure 11-1 on pages 186–187.

This analysis is based on wood, fossil fuel, capital, and labor costs and interest rates prevailing at the time of publication. It also includes estimates for the escalation of these costs throughout the life of the facility. The output shows an after-tax discounted internal rate of return (IRR) of 39.1%, which is well above the hurdle

**Table 11-1.** Sensitivity analysis

| Wood fuel cost, delivered ($/ton) | After-tax discounted IRR |
|---|---|
| $8 | 42.7% |
| $10 | 41.7% |
| $12 | 40.7% |
| $15 | 39.1% |
| $18 | 37.6% |
| $20 | 36.6% |

rate (8–10%) generally required for an investment to be considered profitable.

A sensitivity analysis was performed to determine the after-tax discounted IRR over a range of wood fuel costs. A range of prices was selected ($8.00 to $20.00/ton) for wood fuel, and the corresponding returns are shown in Table 11-1. Even in the case of the highest wood fuel cost ($20/ton), the project is still shown to be profitable.

# CHAPTER 12

# BIOMASS FUEL PROCESSING ROUTES AND ECONOMICS

## INTRODUCTION

The nonforest products industry is confronted with a wide variety of options when it decides to switch to wood fuel. Should wood residue be purchased as fuel, or are whole-tree chips preferred? What type of boiler and burner are indicated? Is drying or pelletizing economically attractive? This quantifies cost to the user for each biomass processing route as the cost of energy (typically as steam) in dollars per million Btus delivered. This work was originally reported in June 1980 [1], and for a more complete treatment this report is recommended.

Emphasis is on direct combustion of wood in boilers, as this is the most common commercial use of biomass energy. Other less common application methods, such as wood densification and gasification, are dealt with, and rough cost estimates are provided.

## ECONOMIC ANALYSIS

The prime ingredients in the economics of fuel systems are capital, operating, and fuel costs. Each company uses a different method to decide on the viability of an investment, dictated by their method of financing, tax liabilities, and so on. It is difficult to compare fossil-fueled systems using gas or oil with solid-fueled systems using biomass or coal, as the capital investment and fuel costs are radically different. For these reasons, uniform annual cost and payback period are used in this chapter. The uniform annual cost is divided by the useful heat delivered in order to allow comparison of options over a range of sizes.

Net present worth, another widely used investment analysis method, is not presented, but can be developed using the costs contained in this chapter. Net present worth has a weakness in analyzing fuel systems, in that fuel costs for the proposed system and its alternate must be projected over the lifetime of the equipment. The rate of inflation must be factored in for maintenance, taxes, and so on, all of which are difficult to predict in the near term and impossible to predict 20 years into the future.

Economic decisions are made, however, and a frequently used rule of thumb is a three-year payback period based on first-year economics. For periods longer than this, risk goes up and the project is both less likely to be built and less likely to have positive economics. It is difficult to predict the future load for the system, and recessions and lower product demand due to competition and reduced load and can adversely effect returns. These factors reinforce the use of no more than a 3 year payback as a decision point.

The technique for calculation of system cost is the same for all systems. The total capital required for the system is computed by adding the cost of the components. Tax breaks vary with time and politics, and an assumed investment tax credit of 20% for biomass systems and 10% for coal, gas, and oil systems is utilized in this analysis. The difference between the total capital cost and the investment tax credit is the financed capital cost. This is the amount of money that the user must borrow. The annual cost of capital is based on a 15% interest rate and a loan period of 25 years. The maintenance cost, tax, and insurance are based on a fixed percentage of the total capital cost developed from existing installations. The operating costs are estimated from existing installations and vary with size and type of system. The fuel cost is computed using the fuel consumption rates and the unit cost of the various fuels.

On a practical level, a firm buying a biomass system must also look at cash flow before, during, and after construction to assess what funds are needed, when they will be needed, and the cost of financing the capital required until the process is up and running and producing positive cash flow.

## PROCESSING ROUTES

Biomass can be used in a variety of ways. The simplest and most widely used processing route is the direct combustion of green (50% moisture content) wood chips for producing steam. This route has a minimum of intermediate steps, composed of unloading, storage, conveying, and burning. More complex routes include sizing, drying, densification, gasification, and liquefaction before combustion. Direct combustion is the most viable route, and will be dealt with in detail. Other operations which show future promise are presented in a condensed form. In Chapter 13, a complete graphic of wood processing routes is presented.

## BIOMASS-DERIVED FUELS

The transformation of biomass into a more usable form is technically feasible and rapidly approaching economic viability. Included in this category are gasification, liquefaction, and densification. These operations upgrade the form of the fuel by cutting the weight and volume, reducing the cost of transportation and combustion, and allowing the fuel to be used in a wider range of equipment.

Biomass gasification is under development in the United States, Canada, and Europe. Gasifiers fueled with wood and coal were used from the late 1800s through World War II when cheap oil and gas forced them from the marketplace. Their prime market at this time is the retrofit conversion of gas/oil boilers, dryers, and kilns. They can also be used to fuel internal combustion engines or turbines. Gasifiers produce a low-Btu gas that is used onsite and typically contains 150 Btu/ft$^3$ to 300 Btu/ft$^3$.

Wood and ag residue can be used as feedstock for producing ethanol, methanol, pyrolysis oil, and catalytic oil. Ethanol is made by converting cellulose into glucose followed by fermentation. The methanol route uses wood gasification to produce a CO/H$_2$ mixture followed by shift reaction. Pyrolysis oil, char, and gas are produced by thermal decomposition in the absence of air. Catalytic oil is made using alkaline catalysts at high pressures and temperatures. The alcohols can be used for transportation and may command higher prices, whereas the oils would be burned in place of fuel oils.

Densified wood is being produced in the form of wood pellets. This process includes drying, grinding, and producing pellets by forcing the finely ground wood through a die. This fuel can be substituted for coal in existing equipment and may find markets in gasifiers and commercial installations. The current market in the United States is residential and light commercial heating systems. The bigger and most recent market is export and use in Scandinavian utility boilers for Kyoto protocol compliance on greenhouse gases. The production of densified fuels is covered in detail in the following section. In addition to pellets, fuel logs are produced. These are used for residential fireplaces and are not addressed further in this text.

The results of the economic analysis of wood-derived liquid and gas fuels are found in Table 12-1. The total cost of each product is based on capital, operating, and fuel costs for typical plants (a more detailed analysis may be found in [1]).

The cost of the liquid fuels is higher than competing fossil fuels, but the gap is closing rapidly as gasoline and fuel oil prices increase. Wood pellets, more costly than raw wood or coal, compete with natural gas. Low Btu gas is cost competitive with natural gas and less costly than oil; hence, the renewed interest in this technology.

## DENSIFIED WOOD AND WOOD PELLET PRODUCTION

Densification of wood has been done in a variety of ways. For industrial use, it has focused on using dies to produce pellets based on friction producing heat; the heat causes lignin that is part of

**Table 12-1.** Cost to produce wood-derived liquid and gas fuels

| Fuel | Cost/unit | Cost/million Btus |
|------|-----------|-------------------|
| Ethanol | $2.75/gal | $32.60* |
| Methanol | $1.76/gal | $27.30* |
| Pyrolysis Oil | $1.26/gal | $10.86* |
| Pyrolysis Char | $239/ton | $9.07* |
| Catalytic Oil | $2.84/gal | $19.11* |
| Low Btu Gas | $0.82/MCF | $5.40* |
| Wood Pellets | $150/ton[†] | $9.80 |

*Costs inflated 1984–2008 via CPI factor of 2.1.
[†]Current market price, 2008.

wood to act as a natural binder. Other approaches include making logs via extruders, and using wax and other binders rather than depending on heat and lignin. One of the authors (Grant Curtis) of the 1984 *Industrial Wood Energy Handbook* also experimented with use of natural cellulosic binders for charcoal briquetting, and the editor worked on marketing the process to reduce deforestation in developing countries by allowing the use of previously wasted charcoal fines.

Wood pellets were first demonstrated in the early 1980s. Pellet fuel (see Figure 12-1) is manufactured using technology similar to that for grain-based animal feeds. Biomass pellets can be generated from wood chips; wood wastes, including sawdust and shavings; and even agricultural waste. Initially, pellet production was driven by residential demand that followed the introduction of commercially available residential wood pellet stoves in 1984, but recent concerns over greenhouse gas (GHG) contributions to global warming has resulted in significant increases in electrical generation with wood pellets.

Pellets are an attractive fuel due to their high density and heating value and low moisture content. The small size and uniform shape make pellets ideal for automatic feeding to a burner through mechanical auger or pneumatic conveying. Primary applications are residential wood furnaces as a method to offset high oil

**Figure 12-1.** Wood pellets.

and natural gas prices and as industrial or power generation fuel to address carbon taxes and renewable energy and greenhouse gas emission regulations.

Table 12-2 presents a comparison between densified wood pellets and other forms of wood fuel. Pellets, because the feedstock is dried prior to densification, have lower moisture content than sawdust, chips, and even dried material like planer shavings. The densified nature of pellets is demonstrated by their high bulk density of 35 lb/ft$^3$, well above that of sawdust and chips and significantly higher than that for dry material such as planer shavings.

There is room for other concepts, and while not yet commercial, the editor has developed a process for production of low-voidage, super-dense dry wood fuel. Densities are in the range of 45 lb/ft$^3$, and the product has better handling and shipping properties, is dust free, and has better combustion properties and lower shipping costs than do pellets.

The high density and low moisture content make wood pellets an attractive fuel because of high energy content. For example, a cubic foot of green sawdust contains 80,000 Btu, dry sawdust contains 80,040, dry planer shavings contains 41,760, and wood pellets contain 252,000 Btu. The favorable density and Btu content combine to give pellets a large energy content advantage over comparable forms of wood energy. The energy concentration advantage make pellets the preferred choice in applications where wood fuels must be transported significant distances. The dry pellets are also biologically stable, and are not prone to the spontaneous combustion that occurs with wet wood waste. They do produce a fine dust during handling, and this is a potential explosion problem.

**Table 12-2.** Wood fuel: average properties

| Wood fuel | Moisture content, wet basis | Heating value (Btu/lb) | Bulk density (lbs/ft$^3$) |
|---|---|---|---|
| Whole-tree chips | 50% | 4,000 | 24 |
| Green sawdust | 50% | 4,000 | 20 |
| Dry planer shavings | 13% | 6,960 | 6 |
| Dry sawdust | 13% | 6,960 | 12 |
| Wood pellets | 10% | 7500–8000 | 35 |

## Pellet Production Process

Although wood pellets possess advantages due to uniform size, high density, and high energy content, the process to generate pellets is relatively complex and capital intensive.

Figure 12-2 shows the eight major steps in the production of wood pellets. The first six steps are the pellet formation process, and the final two steps address product quality and preparation for delivery. Prior to beginning the production process, the pellet raw material must be received, stored, and retrieved from storage. Several days of raw material supply in the form of sawdust and wood chips can be stored on-site in an open pile.

Raw material is recovered from the storage pile with a front-end loader and fed into the wood handling system for processing.

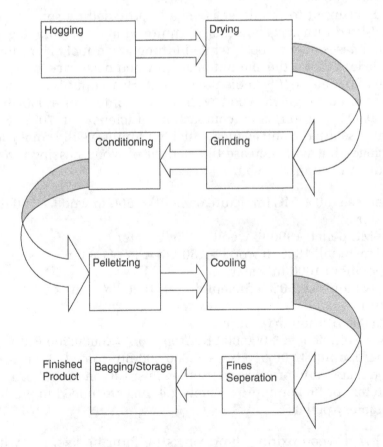

**Figure 12-2.** Wood pellet product: flow diagram.

Although the recovery of wood from storage is not automatic, after it is recovered the material is fed through the process automatically. The initial step following recovery is screening. Wood pieces ½" or smaller fall through a vibrating screen while larger pieces are conveyed to the end of the screen and fall into a wood hog for size reduction. Hog discharge is reduced to the proper size and can be mixed with screened material.

Material of the proper size is conveyed by belt conveyor to a wood dryer. Different types of dryers could be used, and common designs include single-pass and triple-pass rotary drum dryers. Rotary drum dryers are favored because of their high throughput and reasonable cost. Dryers are designed to reduce the moisture content of the feed material from 50% at input down to 10% at discharge. The thermal efficiency of a convective dryer is generally around 50%.

Heat input to the dryer is usually provided by some type of wood-fired furnace. The furnace employed must be able to generate hot gases above 800°F while burning green fuel. The calculation below shows the amount of green wood necessary to produce a ton of pellets at 10% moisture content. Heat input is assumed to be supplied by green wood with a heat content of 4,000 Btu/lb (8,000,000 Btu/ton) and combustion efficiency of 70%. Some plants use dried wood as fuel, which will improve the combustion efficiency but will increase the amount of wood passing through the dryer to supply the burner system.

Pellet feed: 50% MC (moisture content) = 500 lb wood + 500 lb water
10% MC pellet: 1800 lb wood + 200 lb water
Feed material: 1800 lb wood + 1800 lb water (1.8 ton of feed)
Evaporation: 1600 lb/water
Energy required (50% efficient dryer): 1600 lb/water × 1000 Btu/lb ÷ 0.5
3,200,000 Btu/ton dryer input
Furnace efficiency: 8,000,000 Btu/ton × 0.7 = 5,600,000 Btu
Furnace input: 3,200,000 Btu ÷ 5,600,000 Btu/ton = 0.6 ton
Wood material to produce one ton pellets: 1.8 ton for pellet + 0.6 ton for fuel or approximately 2.4 tons green feed to produce 1 ton dry pellets

Dried wood exiting the dryer is uniform in size, but with a maximum dimension of ½" it is still too large for pelletizing. Prior to entering the pellet mill, the wood chips are reduced to a uniform powder consistency capable of passing through a ⅛″ screen (about 3 mm maximum dimension) in a hammermill (Figure 12-3). Depending on the pellet mill design and final pellet diameter, screens larger or smaller than ⅛″ may be used.

After the wood feed is reduced to a uniform, fine consistency in the hammermill, the wood is prepared for pelletizing in the conditioner. In the conditioner vessel, steam is injected to heat and lubricate wood entering the pellet mill. Figure 12-4 shows an external view of a typical pellet mill. This design includes an integral conditioner on top of the mill. After it leaves the conditioner, a wood feeder conveys raw material into the center of the mill where rollers extrude the feedstock through a die to form a cylindrical pellet.

Figure 12-5 shows the internal parts of a pellet press. Rollers force wood flour through a die, extruding it and forming a cylindrical pellet of the desired diameter. The pellet mill often includes a knife to cut the pellets off at the desired length.

After the pellets are formed, they are fed into a cooler to reduce their temperature. The extrusion process involves significant friction between the wood and die as well as compression of the wood. Both of these factors increase the temperature of the pellets to from 200–250°F. In the cooler, ambient air is pulled up through the pellet bed to carry heat away. Coolers are sized to discharge

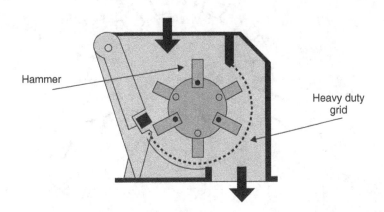

**Figure 12-3.** Hammermill internal parts. (Courtesy of "Wood Waste Recovery: Size Reduction Technology," NIST MEP Recycling Technology Assistance Program, Report CDL-97-3, prepared by CPM Consultants, Seattle, WA, December, 1997.)

**Figure 12-4.** Andritz 26 LM pellet mill. (Used with permission, Andritz Sprout Bauer [2].)

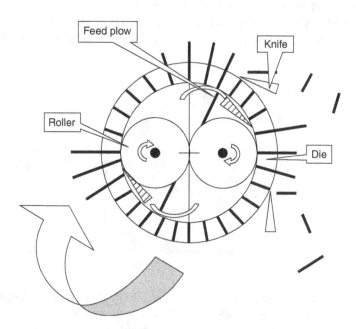

**Figure 12-5.** Schematic of pelleting process.

pellets at approximately 10°F above the ambient temperature. The design air flow rate is approximately 1000 cfm per ton of pellets/hr.

The final two steps in the process are included to improve product quality and enhance distribution. For fines separation, the pellets are screened to separate and collect dust. Then the pellets are bagged for sale to distributors or placed in storage silos for large volume sales via truck or rail.

The typical arrangement of a wood pellet mill is presented in Figure 12-6. The schematic shows the location of the major process components including hog, dryer, hammermill, pellet mill, cooler, fines separator, and bagger. A component not shown in Figure 12-6 is the conditioner; however, in many cases the conditioner is an integral part of the pellet mill. Figure 12-5, a detailed photograph of a pellet mill, shows the position of the conditioner.

## Wood Pellet Industry

In 1984, following the introduction of residential wood pellet stoves, there were two domestic manufacturers of pellets, both located in the Pacific Northwest. Following the growth in pellet demand by the industrial and power generation sectors, the number

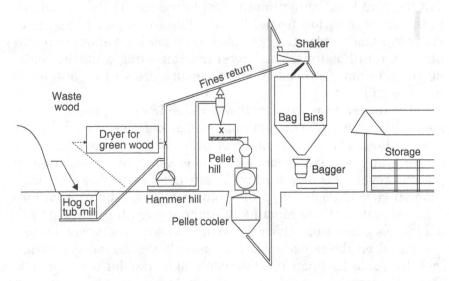

**Figure 12-6.** Typical wood pellet plant layout [3].

of pellet plants grew to over 80 wood pellet production plants in North American by 2007, producing in excess of 1.1 million tons of fuel each year. Pellet manufacturers are represented by an active trade association, the Pellet Fuels Institute, which can be found on the internet at www.pelletheat.org.

The European Commission Directorate on General Energy and Transportation financed a study in 2000 of the wood pellet market in Europe [4]. The study examined wood pellets in four countries: Sweden, Norway, Germany, and Austria. The countries were selected because their pellet market was most highly developed, and their mature economies will stimulate market development in neighboring countries.

The study found that wood pellets have an enormous potential as a biomass fuel although the current market share (2000) was small. Favorable characteristics include raw material availability, comparatively high energy density, applicable to automatic firing systems, and competitively priced, assuming certain energy policy considerations are fulfilled. Given these conditions, the study indicated that wood pellets could become an attractive alternative to fuel oil or electricity for residential heating applications.

Wood pellet exports grew substantially between 1999 and 2005, driven primarily by demand in Europe, as shown in Figure 12-7. Much of the increased demand for pellets is the result of European countries attempting to comply with the Kyoto Protocol and the mandated reduction in greenhouse gas (GHG) emissions. In the six-year period from 1999 to 2005, wood pellet exports in the world market effectively tripled, growing from about two million to six million tons. The largest five importing countries, Italy, Belgium, Denmark, Sweden, and Germany, are all part of the European Union.

The demand for wood pellets will be driven by legal requirements, the price of conventional fuels, and the availability of wood pellets. The primary markets for wood pellets are residential heating and electrical generation. Table 12-3 shows the aggressive plans of several EU countries with respect to renewable electricity generation. Although all of the countries listed show some biomass generation, there remains substantial opportunity for growth of biomass generation, given the required percentages remaining.

Based on the target goals for renewable electrical generation in 2010, Table 12-4 presents the potential market for wood pellets. The table assumes that each listed country satisfies the expected

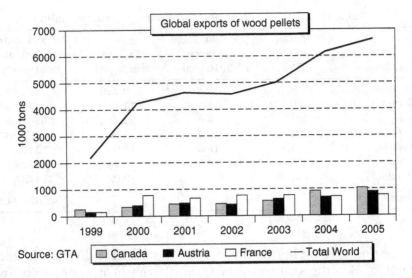

**Figure 12-7.** Global pellet imports. (Courtesy of Wood Resource Quarterly, Global Wood Fiber Market Update, Wood Resources International, Seattle, WA, July, 2006 [5].)

renewable generation, and all electricity is generated at 33% thermal efficiency with wood pellets. The listed countries have other options for renewable generation, including wind, geothermal, and hydro, but because much of the indigenous renewable resources are already being utilized, addition renewable supply may depend on imported wood pellets. Assuming that only 25% of the projected renewable energy is supplied by imported wood pellets, the European electrical generation market demand could increase 10,600,000 tons per year by 2010.

**Table 12-3.** Electricity generation profile, 2005 [6]

| Country | % Renewable | % Biomass | Target %, 2010 | Remaining % |
|---|---|---|---|---|
| Denmark | 28.2 | 10.6 | 29.0 | 0.8 |
| Germany | 10.6 | 2.7 | 12.5 | 1.9 |
| Italy | 14.4 | 1.7 | 25.0 | 10.6 |
| Netherlands | 7.5 | 5.7 | 9.0 | 1.5 |
| Austria | 59.8 | 3.1 | 78.1 | 18.3 |
| Finland | 26.9 | 11.0 | 31.5 | 4.6 |
| Sweden | 54.3 | 5.5 | 60.0 | 5.7 |

**Table 12-4.** Potential biomass market for electrical generation

| Country | Total generation (TWh) | % Renewable to reach 2010 goal | TWh to reach goal | Pellets required to generate goal (tons/yr) |
|---|---|---|---|---|
| Denmark | 36.3 | 0.8 | 0.290 | 176,671 |
| Germany | 620 | 1.9 | 11.7 | 7,170,094 |
| Italy | 303 | 10.6 | 32.1 | 19,584,846 |
| Netherlands | 100 | 1.5 | 1.50 | 914,383 |
| Austria | 65.7 | 18.3 | 12.0 | 7,314,522 |
| Finland | 70.5 | 4.6 | 3.24 | 1,972,951 |
| Sweden | 158 | 5.7 | 9.02 | 5,492,872 |
| Total | | | | 42,626,343 |

Another growing market for wood pellets in Europe is home heating. Heating systems vary from district heat systems in Sweden to individual wood pellet stoves common in Austria, Germany, and Denmark. Sale of wood pellet heating equipment has grown on average between 30–50% per year during the last decade [7]. As heating oil prices rise, the only limit to increased wood pellet demand is the available supply.

Data shows that global pellet imports tripled in the six years from 1999 to 2005. This is equivalent to a growth of about 50% per year. The expanded emphasis on renewable electrical generation and growth in excess of 30% per year in the home heating market both contribute to the expected European market growth. Worldwide pellet production was 7 million tons in 2007 according to the Wood Pellet Association of Canada. The pellet market continues to grow both domestically and globally, and European pellet demand is expected to increase from 5 million tons in 2006 to 13 million tons in 2010, a 260% growth rate [8].

The wood pellet industry is characterized by a small number of large manufacturers (capacities greater than 50,000 tpy) and a large number of small manufacturers (capacities less than 50,000 tpy). Prior to the construction of the large pellet plants beginning in 2007, there were approximately 40 plants operating in the United States with a combined output of 900,000 tpy [9]. In Europe, there are more than 200 pellet mills with production capacities from 5 to 15,000 metric tons per year. The output of the European mills is limited by the availability of raw material.

The growing demand for wood pellets has resulted in the entry of several large competitors into the market located primarily in

North America. The availability of large amounts of raw material in North America has resulted in larger pellet plants than in Europe. North American mills range from 50,000 to over 600,000 tpy.

When determining the relative competitive position of wood pellet plants in the marketplace, several factors must be considered:

1. Plant investment
2. Raw material cost
3. Shipping cost

Table 12-5 summarizes the pellet output, plant investment, and investment per tpy of output for several southeastern pellet mills. The range in investment cost per ton of output is $160 to $197. The investment costs are approximate based on figures reported in newspapers and trade publications.

The reported investments for new pellet plants are comparable. It is worth noting that inflation may be reflected in the cost figures below. The lowest investment ratio is for Dixie Pellets in Selma, Alabama, which was completed in mid 2007. Green Circle and Fram were both completed in late 2007, and they have investment ratios that differ by 7%. Finally, the plant planned for completion in mid-2008, Dixie Pellets in Jackson, Alabama, has the highest investment ratio of $197/tpy.

In the span of 1½ years, the investment ratio for wood pellet plants in the southeast has risen 23%. This may indicate that pellet plants entering the market later will be at a cost disadvantage compared to the earlier ones. Also, these investment figures reveal that there is essentially no economy of scale in pellet production.

The second competitive factor that must be considered is the raw material source and cost. Fram is located in an area adjacent to several large hardwood sawmills and will draw sawmill waste

**Table 12-5.** Southeastern U.S. wood pellet plants

| Plant site | Pellet output (tpy) | Investment ($) | Investment ratio ($/tpy) |
|---|---|---|---|
| Fram Fuels, Baxley, GA | 130,000 | 22,000,000 | $170 |
| Green Circle, Panama City, FL | 550,000 | 100,000,000 | $182 (10) |
| Dixie Pellets, Selma, AL | 500,000 | 80,000,000 | $160 (7) |
| Dixie Pellets, Jackson, AL | 600,000 | 118,000,000 | $197 (7) |

to pelletize and bark to burn from them. Because there are no papermills in the immediate vicinity and Fram is using hardwood instead of softwood, Fram faces no significant competition for raw material. Because hardwood is not used as a raw material for paper, it can only be used as fuel by them. The market price of hardwood waste is about $15/ton.

The raw material used by the other plants is sawmill waste, if available, or harvest waste and selective thinning [10]. Pine chips have higher cost than hardwood sawmill waste because they can be used for fiber in papermaking. Selective thinning from pine forests will have higher costs because one pellet plant has an agreement with the sellers that the material will be utilized for the highest value product, pellets or lumber. Danny Duce, the Green Circle assistant wood procurement manager, said "We will merchandise the timber according to the highest value to the (seller)" [11]. Based on this procurement strategy, pellet plants in Alabama and Florida will be paying from $20 to $30 per ton for raw material.

A final factor to be considered, especially when exporting pellets, is shipping costs. The plants in Alabama and Florida will be shipping from the Gulf, and their shipping costs will be from $5 to $10 per ton higher than plants on the Atlantic because of the longer distances involved.

## Wood Pellet Prices

An important factor in determining the demand for wood pellets is the cost. Because pellets are a refined form of biomass, their retail price is significantly higher than traditional biomass energy sources like split logs, sawdust, or bark. Pellet price depends on the quantity purchased, raw material used in production, and the distance transported.

To determine the selling price for pellets, historical sources were consulted. When determining price estimates for pellets, it is important to consider only pellets sales. Prices for wood chips, sawdust, or bark do not offer a suitable comparison because wood pellets are denser, dryer, and more uniform in size, and have higher energy content than other forms of biomass fuel.

The costs for pellets in Sweden are tracked by the Swedish Energy Agency. In their publication, Prisblad, from early 2008, the pellet cost was $191/ton in 2007 [12]. Pellet cost data from this publication is presented in Table 12-6. Notice that in Sweden, pellets are sold based on heat content. A megawatt hour is equivalent to 3,412,000 Btu. The heat content of wood pellets is approximately 4.7 MWh/ton or 8000 Btu/lb. To convert cost, the value of a Swedish Krona (SEK) is $0.1665. The table shows a price decline in 2005 as a result of the mild winter. Growing usage by electrical generators has created upward pressure on prices of late.

Another reference noted that the price of wood pellets in Europe increased significantly during the winter of 2006 due to raw material shortages. In 2005, the price of pellets for home heating was €180/tonne ($218/ton) but rose quickly to €270/tonne ($295/ton) in 2006 [4]. Although this price is for small quantity deliveries, it shows the upward pressure on prices. During the winter of 2007, a similar shortage of wood pellets caused a price rise domestically. In the northwest United States, the price of pellets was $180/ton prior to the shortage and $195–$200/ton after [13]. Another reference states that bulk deliveries of pellets range from $200 to $230/ton in the United States, depending on region of the country [7]. Although there is significant variation in wood pellet prices both regionally and over time, growing demand resulting from higher petroleum prices and the conversion to renewable energy for electrical generation has caused an escalation in wood pellet prices.

With an energy content of 7500 Btu/lb or 15 million Btu/ton, the energy cost per million Btus is directly related to the price of fuel. Table 12-7 presents the energy cost per million Btu for the pellet prices presented in the previous paragraph. Pellets can no longer be purchased for $100/ton because this is just slightly more than the cost of production. Pellets at $150/ton converts to an energy cost of $10 per million Btu. When the cost of pellets is $200/ton, the equivalent energy cost is $13.33 per million Btu. In many locations, this cost is lower than what homeowners are pay-

**Table 12-6.** Swedish Energy Agency price data

| Material, Units | 2004 | 2005 | 2006 | 2007 |
|---|---|---|---|---|
| Pellets, SEK/MWh | 206 | 204 | 211 | 244 |
| Pellets, $/ton | 161 | 160 | 165 | 191 |
| Chips, SEK/MWH | 125 | 121 | 119 | 128 |
| Chips, $/ton | 98 | 95 | 93 | 100 |

**Table 12-7.** Unit energy cost for different pellet prices

| Pellet cost ($/ton) | Energy cost ($ per million Btu) |
|---|---|
| 100 | 6.67 |
| 150 | 10.00 |
| 180 | 12.00 |
| 200 | 13.33 |
| 250 | 16.67 |
| 290 | 19.33 |

ing for fuel oil or propane. If pellets cost $290/ton, the unit energy cost is approaching $20 per million Btu. Only if conventional fuels are not available or incur some added penalty, like a carbon tax, would pellets be the fuel of choice at this price for pellets.

These pellet prices are generally competitive with natural gas and fuel oil on a $/MM Btu basis (as shown in Table 1-1).

## Wood Pellet Summary

Recent policy and market changes have stimulated increased demand for wood pellets. The growing application of pellets for residential heating and green electrical power is a major contributor to this increased demand. Sweden, Canada, and the United States are the largest producers of pellets, with a combined production of 3.85 million tons [14]. Sweden, in addition to being a large producer, is also the largest user of wood pellets. Other large markets are also European countries, including Austria, Italy, Germany, the Netherlands, Denmark, and Belgium.

Globalization is rapidly occurring in the pellet market and trade is becoming a key feature. A number of pellet producers are selling in Europe as a result of the maturing market, but emerging markets are opening up in both Latin America and Asia.

Although wood fiber will always be demanded by the paper and building products industries, higher energy prices will draw a certain amount of waste wood to pellet production. The global quantity of wood waste available for pellet manufacturing is about 2800 million cubic feet, which could be converted into 14 million tons of pellets. If this occurs, the wood pellet production would triple from the production in 2006 [14].

# DIRECT COMBUSTION FOR STEAM GENERATION

Wood presently supplies approximately 2% of U.S. energy needs, with many projections showing a possible contribution of 7% possible on a renewable basis, and is the biggest single biomass source. The majority of present use is in producing steam for industrial process heat in pulp and paper mills. Nonforest product industries located in the forested areas of the United States are now turning to this resource. Typical new installations are wood-fired boilers of 50K–200K steam capacity rated at 150 psig saturated steam.

In this section, the capacities of individual boilers and entire boiler plants examined span the ranges of 10,000 to 200,000 lb/hr of steam. Five different boiler sizes are analyzed in this range, using seven different systems. These systems are a wood waste boiler, wood chip boiler, wood pellet boiler, wood fueled fluidized-bed combustor, coal boiler, gas boiler with oil backup, and an oil boiler. The costs of the wood waste gasifier with gas/oil boiler are presented in the summary tables and charts. Five different sizes are analyzed for each of these systems: 10,000 lb/hr steam, 25,000 lb/hr steam, 50,000 lb/hr steam, 100,000 lb/hr steam, and 200,000 lb/hr steam. The systems have been configured for almost completely automatic operation, although some applications may not require this degree of automation.

The technique used to analyze each of the seven systems is the same. For each system, a process diagram is developed that shows the routing of the fuel from arrival at the plant to combustion. The cost of each piece of equipment is identified and entered in a system cost table.

The object of the system cost table is to determine the cost of steam in dollars per million Btus delivered for each system analyzed. These costs are then compared to determine the attractiveness of the various systems. The cost of steam is computed by dividing the total annual cost of a system by the annual amount of heat delivered.

The design factors used in the analysis of each of the systems are presented in Table 12-8. Each of the boilers is assumed to generate 150 psig saturated steam, with inlet feedwater at 220°F and efficiencies as shown. The fuel data presented was representative of heat values and costs in Georgia in the early 1980s. The plants analyzed are assumed to operate 24 hours per day, 345 days per year, and at an average load factor of 70%. This data can be modi-

**Table 12-8.** Design factors

| Boiler operating conditions | |
| --- | --- |
| Operating pressure | = 150 psig saturated |
| Inlet feedwater temperature | = 220°F |

| Efficiencies (%) | |
| --- | --- |
| Waste wood boiler | = 65 |
| Wood chip boiler | = 65 |
| Wood pellet boiler | = 80 |
| Wood fueled fluidized bed combustor | = 68 |
| Wood gasifier | = 85 |
| (with gas boiler) | = 69 |
| Coal boiler | = 83 |
| Gas boiler | = 81 |
| Oil boiler | = 85 |

Fuel Data (1980 $)

| Fuel | Heat value | Unit cost | Raw fuel cost ($ per million Btu) |
| --- | --- | --- | --- |
| Waste wood | 4,250 Btu/lb | $7.00/ton | 0.82 |
| Wood chips | 4,250 Btu/lb | $10.50/ton | 1.24 |
| Wood pellets | 7,650 Btu/lb | $45.00/ton | 2.94 |
| Coal | 13,000 Btu/lb | $35.00/ton | 1.35 |
| Gas | 1,000 Btu/gal | $0.30/$10^5$ Btu | 3.00 |
| #6 Oil | 150,000 Btu/gal | $0.65/gal | 4.33 |

| Plant operation | |
| --- | --- |
| Hours/day | = 24 |
| Days/year | = 345 |
| Load Factor | = 70% |

fied as required for analysis of a particular installation. Hot oil heaters are not included in the evaluation, but are similar to steam boilers in process layout and operation.

## Waste Wood Boiler System

The process diagram of a waste wood boiler system is shown in Figure 12-8. The waste wood is delivered to the plant by truck, unloaded with a hydraulic truck dumper, and either immediately processed into the system or moved to and from outside storage using a front-end loader. The wood enters the automatic system through a drag chain conveyor that feeds a belt conveyor having a

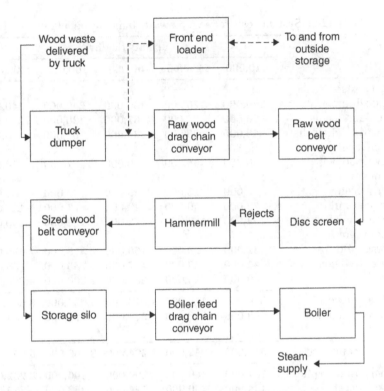

**Figure 12-8.** Process diagram of a waste wood boiler system.

20-foot horizontal section for fuel transfer. The wood waste is processed over a disc screen and the rejects are further reduced in size by a hammermill. The fines are allowed to pass through without size reduction. Whole-tree chips, on the other hand, do not follow this processing route. The sized wood is transported by belt conveyor to a storage silo with sufficient capacity for 3 days of boiler operation. The fuel from the storage silo is transported to the boiler using a drag chain conveyor.

The system cost for waste wood boiler systems is shown in Table 12-9. Since the system utilizes wood, which is a renewable energy source, a 20% investment tax credit is used. The operating cost is a percentage of the total capital cost taken from operating experience. The maintenance cost is assumed to be 5% of the total capital cost, and the tax and insurance are assumed to be 2.5% of the total capital cost. The annual heat delivered is calculated by determining the amount of heat released for a 24-hour day operation, 345 days per year, at a 70% load factor.

**Table 12-9.** System costs for a wood waste boiler system (1982 $)

| | Capacity (lbs/hr) | | | | |
|---|---|---|---|---|---|
| | 10,000 | 25,000 | 50,000 | 100,000 | 200,000 |
| Capital costs | | | | | |
| Truck dumper | 120,000 | 120,000 | 120,000 | 240,000 | 240,000 |
| Front-end loader | 16,000 | 22,000 | 57,000 | 120,000 | 240,000 |
| Raw wood drag chain conveyor | 18,000 | 25,000 | 38,000 | 65,000 | 116,000 |
| Raw wood belt conveyor | 56,000 | 58,000 | 60,000 | 67,000 | 74,000 |
| Disc screen | 7,000 | 13,000 | 23,000 | 37,000 | 66,000 |
| Hammermill | 22,000 | 30,000 | 43,000 | 71,000 | 125,000 |
| Sized wood belt conveyor | 154,000 | 163,000 | 172,000 | 189,000 | 200,000 |
| Storage silo | 32,000 | 79,000 | 158,000 | 315,000 | 630,000 |
| Boiler feed conveyor | 41,000 | 43,000 | 47,000 | 53,000 | 68,000 |
| Boiler | 350,000 | 625,000 | 900,000 | 2,800,000 | 4,800,000 |
| Total capital cost | 816,000 | 1,178,000 | 1,618,000 | 3,957,000 | 6,319,000 |
| Less investment tax credit | 163,000 | 236,000 | 324,000 | 791,000 | 1,264,000 |
| Financed capital cost | 653,000 | 942,000 | 1,294,000 | 3,166,000 | 5,055,000 |
| Annual cost of capital | 101,000 | 146,000 | 200,000 | 490,000 | 782,000 |
| Operating cost | 134,000 | 150,000 | 194,000 | 352,000 | 379,000 |
| Maintenance cost | 41,000 | 59,000 | 81,000 | 198,000 | 316,000 |
| Tax and insurance | 20,000 | 29,000 | 40,000 | 99,000 | 158,000 |
| Fuel cost | 74,000 | 184,000 | 370,000 | 741,000 | 1,480,000 |
| Total annual cost | 370,000 | 568,000 | 885,000 | 1,880,000 | 3,115,000 |
| Annual heat delivered ($10^6$ Btu) | 59,000 | 146,000 | 292,000 | 585,000 | 1,168,000 |
| Cost per million Btus delivered | $6.27 | $3.89 | $3.03 | $3.21 | $2.67 |

## Wood Chip Boiler System

The process diagram for a wood chip boiler system is similar to that shown in Figure 12-8; however, the disc screen, hammermill, and raw-wood belt conveyor are eliminated. Since the wood chips already have a somewhat uniform size, no size reduction equipment is required and the fuel is transported immediately to a storage silo with a capacity for 3 days of boiler operation. Wood is transported from the storage silo to the boiler using a drag chain conveyor.

The costs for this system and the other systems that follow were computed in a similar manner to those for the waste wood system. Total capital costs and annual fuel costs for the systems analyzed are presented in Figures 12-9 and 12-10 in 2008 dollars.

## Wood Pellet Boiler System

Wood pellets are dumped directly into a bin that feeds a belt conveyor. An advantage of pellets is their low moisture content. Since it is not desirable to add moisture to the pellets, no outside storage is utilized. A drag chain conveyor would tend to break the pellets; therefore, a belt conveyor is used to feed the storage silo and then to feed the boiler.

## Wood Fueled Fluidized-Bed Combustor System to Produce Hot Gas

Fuel for the system (waste wood) is delivered to the plant by truck, unloaded by a hydraulic truck dumper, and transported to and from outside storage by a front-end loader. The wood enters the system through a drag chain conveyor that feeds a belt conveyor. A disc screen is again utilized to separate oversized pieces of

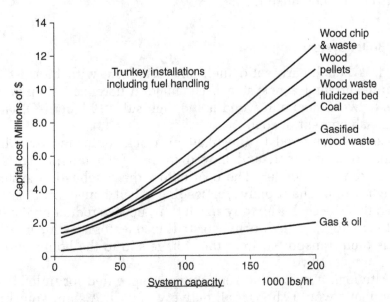

**Figure 12-9.** Capital cost of steam generating system (2008 $).

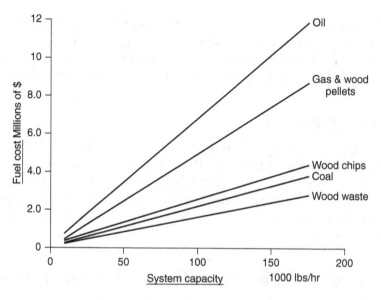

**Figure 12-10.** Annual fuel cost for steam generating systems (2008 $).

wood, and a hammermill is used for size reduction. A belt convey-
or transports the wood to a storage silo, and a drag chain or screw
conveyor is used to transport the wood from the storage silo to the
fluidized-bed combustor.

## Coal Boiler System

Coal is a benchmark solid fuel, and competes with biomass in
medium and large size systems. The major difference in fuel properties between coal and wood is the high sulfur, higher $NO_x$, and
higher ash content of coal.

Coal is delivered to the plant by truck, unloaded by a hydraulic truck dumper, and moved to and from outside storage
with a front-end loader. The coal enters the combustion system
through a drag chain conveyor feeding a belt conveyor. It is assumed that the coal is already sized; thus, no size reduction equipment is required. A belt conveyor will feed a storage silo, and the
coal is then transported from the storage silo to the boiler with a
drag chain conveyor.

Although a hydraulic truck dump is specified for unloading,
many plants would choose rail delivery of coal. A short spur line
with bottom unloading via a trestle would cost approximately

$235,000. Deleting the cost of the truck dump, this system would add $0.03 per million Btus delivered. Long spur lines might increase this cost considerably, and the choice of rail or truck delivery would depend on availability of rail service, size of boiler, and distance to coal mines.

The capital and operating costs of $SO_2$ scrubbing equipment is not included. With smaller boilers in attainment areas using low sulfur coal (less than 2% sulfur), scrubbing may not be required. The addition of $SO_2$ scrubbers would add approximately $5 per lb/hr of steam capacity to the capital cost of the boiler plant. The annual cost for this capital, plus maintenance, would add approximately $0.25 per million Btus delivered to the cost of a coal-fired steam system.

## Gas Boiler System

Due to the possibility of a curtailment of natural gas supplies to an industrial boiler, the system is designed to fire either gas or oil. The gas is supplied to the boiler directly from a pipeline; oil is delivered to the plant by truck. An outside oil storage facility is equipped with a transfer pump, which pumps the oil to a day storage tank that, in turn, feeds the boiler.

The outside oil storage tank is sized for a 7-day emergency supply of oil; its pumps are designed to fill the day tank in one hour. The day tank and outside storage tank are equipped with heaters for the #6 fuel oil.

## Oil Boiler System

Oil is delivered to the plant by truck, stored in an outside storage tank, and pumped by transfer pump to the day tank. The day tank transfers oil directly to the boiler with pumps.

The outside oil storage tank is designed for a capacity providing 30 days of boiler operation. The oil transfer pumps are designed with a capacity to transfer oil to the day tank in one hour; both day tank and outside storage tank have heaters for the #6 fuel oil.

## Wood Waste Gasifier with Gas/Oil Boiler

An updraft gasifier used to retrofit a gas/oil boiler is the basis for this system. It is fueled with whole-tree chips and is close-coupled to the boiler. The steam costs in dollars per million Btus delivered for different fuels are presented in Table 12-10 for comparison with other systems.

## SUMMARY

### Capital Costs

The capital cost of any steam generating system can be separated into *boiler cost* and *fuel system cost*. The latter can be found by subtracting the boiler cost from the total cost. Comparison of the cost data and the efficiency data indicates that boilers for gas and oil systems are efficient and low in cost, and fuel system costs are minimal. In contrast, solid fuel equipment is less efficient, particularly when burning high-moisture-content fuels. The boiler cost and extensive solids handling and storage systems require a large capital investment. Solid fuel equipment is also considerably larger in size.

Figure 12-9 presents capital cost curves for boiler systems, including fuel handling from 10,000 to 200,000 lb/hr steam for gas, liquid, and solid fuels. Solid-fueled boiler systems are approximately nine times the cost of gas/oil systems. Gasifier retrofitting of existing boilers is somewhat less—approximately five times the cost of gas/oil systems.

### Fuel Costs

Figure 12-10 summarizes the fuel cost for the same systems, based on a 345-day-per-year operation and a 70% load factor. Oil is the most costly fuel (#6 @ $0.65 per gallon), followed by natural gas, wood pellets, wood chips, coal (less than 2% sulfur), and wood waste. It is interesting to note that an oil-fired boiler of 30,000 lb/hr capacity with an initial cost of $210,000 will consume $1 million worth of fuel oil in its first year of operation. The annual fuel cost for most solid fuel systems is about one-third of that for gas or oil systems.

The capital cost and fuel costs of the eight steam production systems are inversely proportional; that is, the cheapest fuel requires the most costly handling and burning systems.

### Steam Costs

The primary application in industry for wood fuel and competing fossil fuels is steam production. Using the costs for raw fuel and plant investment covered previously and adding operation costs as a percentage of plant investment, the total cost of producing steam can be found. This cost can be used to compare the economic merits of wood- and fossil-fueled systems for boiler replacement or plant expansion.

The cost of steam for the individual systems is summarized in Table 12-10 and Figure 12-11. Cost is presented in dollars per million Btus delivered of 150 psig saturated steam. This data indicates that the cost of steam goes down with increasing boiler size for all fuels, with the minor exception of a slight jump in price from package boilers to the larger field-erected type over 50,000 lb/hr steam capacity. For the smallest boilers (10,000 lb/hr capacity), natural gas is the least costly system. However, interruptions and rising gas prices may change this position in the future. Coal is more costly, followed by oil, wood waste, wood chips, wood waste for a fluidized-bed combustor, and wood pellet boilers, in order of increasing cost. This analysis points out the fact that solid fuels are expensive to use for small industrial applications.

The 25,000 lb/hr size range is perhaps more representative of industrial boiler sizes. In this size, wood waste is the lowest cost, followed by coal, wood chips, and gas. Natural gas is more costly, followed by wood pellets and fuel oil.

The cost of wood gasifiers retrofitted to existing gas/oil boilers is of interest, although the economic analysis is complicated by the "existing" boiler, which may already be fully depreciated. In

**Table 12-10.** Cost of steam (dollars per million Btus delivered)

|  | System rating | | | | |
|---|---|---|---|---|---|
|  | 10,000 | 25,000 | 50,000 | 100,000 | 200,000 |
| Wood waste boiler | 6.27 | 3.89 | 3.03 | 3.21 | 2.67 |
| Wood chip boiler | 6.37 | 4.29 | 3.53 | 3.76 | 3.30 |
| Wood pellet boiler | 7.96 | 5.91 | 5.10 | 5.29 | 4.90 |
| Wood waste fluidized-bed combustion | 7.86 | 4.61 | 3.45 | 3.06 | 2.68 |
| Coal boiler | 5.76 | 4.05 | 3.44 | 2.98 | 2.69 |
| Gas/oil boiler | 4.61 | 4.34 | 4.16 | 4.05 | 3.99 |
| Oil boiler | 6.12 | 5.80 | 5.61 | 5.45 | 5.38 |
| Wood waste gasifier on gas/oil boiler | 4.54 | 3.10 | 2.50 | 2.22 | 2.01 |

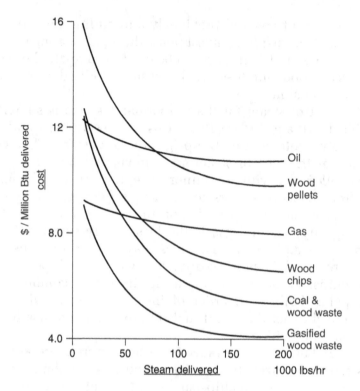

**Figure 12-11.** Total cost of steam.

order to keep the analysis on an equal basis, the cost for delivered steam from a gasifier retrofit system includes the amortization, operation, and maintenance costs of a gas boiler. These costs, based on waste wood, are the least of any system in any size range. Using wood chips as fuel, the cost per million Btu delivered as steam increases $0.61, making it the least costly option, with the sole exception of the natural gas boiler at 10,000 lb/hr capacity. This economic advantage is countered by the lack of dependable hardware at present. Over 30 groups in the United States and Canada are developing wood gasifiers and availability problems may be alleviated in the next few years.

The economics of firing boilers with dry, densified wood pellets are competitive with fuel oil in the 50,000 lb/hr size range and above; below 50,000 lb/hr wood pellets are the most costly fuel. Cost reductions in pelletization technology or the emergence of compact, low-cost gasifiers to convert pellets into a gas compatible with existing gas/oil equipment may improve their competitive

**Table 12-11.** Payback period for steam generating systems (compared to conventional oil boiler)

| Capacity (lb/hr) | Payback period (years) | | |
|---|---|---|---|
| | Waste | Chips | Pellets |
| 10,000 | Oil cheaper | Oil cheaper | Oil cheaper |
| 25,000 | 3.52 | 4.00 | Oil cheaper |
| 50,000 | 1.77 | 1.99 | 6.90 |
| 100,000 | 2.63 | 3.30 | 27.90* |
| 200,000 | 1.69 | 2.20 | 8.20 |

*Due to rising capital cost for field-erected boilers.

position. The solid fuel systems, with larger capital costs, will be more sensitive to increases in loan interest rates than oil and gas systems.

Table 12-11 shows the payback period for wood waste, chip, and pellet fuel systems as compared to a conventional oil-fired boiler. The cost data computed indicates that oil is a cheaper system in the 10,000 lb/hr size range. The cost of the wood systems could be reduced by removing some of the automation and storage designed into the system. Wood waste and chips will pay off within the five-year period, which is generally considered to be the limit for industrial applications. Pellets are not competitive in the smaller capacities and have unacceptably long payback periods in the larger sizes. The jump between 50,000 and 100,000 lbs/hr is due to the increased capital cost of a field-erected versus a packaged boiler. As capacities increase above 100,000 lbs/hr, the boiler costs and the resulting payback periods decrease due to economies of scale.

## REFERENCES

[1]   McGowan, Thomas F., Walsh, James L., Jr., *Wood Fuel Processing*, Volume III, Project A-2400, U.S. Department of Energy Contract, DE-FGO5-79ET23076, Georgia Institute of Technology, June 1980.
[2]   Andritz Sprout Bauer, Pellet Mill Division, Muncy, PA, www.andritzsproutbauer.com.
[3]   Bob Massengill, www.pelletsystemsconsulting.com.
[4]   "Woodpellets in Europe: State of the Art, Technologies, Activities and Markets," Thermie B DIS/2043/98-AT, Industrial Network on Wood Fuels, January 2000.

[5] "Global Wood Fiber Market Update: 1st Quarter 2006," *Wood Resource Quarterly,* Seattle, WA, July 2006.

[6] "Energy and Transport in Figures, Part 2 Energy," European Union, European Commission Directorate—General Energy and Transportation, Brussels, 2007.

[7] "Shortage of Wood Pellets is Causing Heating Woes," *The World,* Coos Bay, Oregon, January 17, 2007.

[8] "Wood Pellet Cost and Availability," *The Encyclopedia of Alternative Energy and Sustainable Living,* www.davidarling.info/encyclopedia/AEmain.html.

[9] www.maerskline.com/link/?page=brochure&path=/our_services/containers/dry.

[10] Jeff Amy, "Timber Owners Could Reap Profits from Pellet Mill," *Press-Register,* Selma, AL, February 2008.

[11] "Green Circle Lands Timberland Contracts: $100 Million Plant Needs Ample Sources of Wood Pellets," *Panama City News Herald,* December 29 2007.

[12] Prisblad för biobränslen, torv m.m., Energygimyndighenten, Nr 1/2008.

[13] Christine Rakos, "Time for Stability: An Update on International Wood Pellet Markets," *Renewable Energy World,* January/February 2008.

[14] Malgorzata Peksa-Blanchard, "Global Wood Pellet Markets and Industry: Policy Drivers, Market Status and Raw Material," IEA Bioenergy Task Force 40, November 2007.

CHAPTER *13*

# BIOMASS FUEL
# PROCESSING NETWORK

A number of processing routes are available for biomass. These multiple paths can be confusing to the newcomer, and without adequate explanation, can give the impression that the use of biomass for fuel is fraught with difficulties. But it all boils down to this: Do as little fuel preparation as possible to get the biomass converted to heat or end product with a particular type of equipment, *and do no more.*

To help clarify the situation, a Biomass Energy Processing Network (Figure 13-1) has been developed. The network is designed to serve as a road map for showing the relationship of individual processing steps to the overall conversion of biomass to energy.

The network has three groups of activities, starting with (1) biomass processing at the source, (2) fuel processing, and, finally, (3) thermal and or electric conversion. The network shows the various routes involved in each phase, based on the most used biomass resource, sawmill waste and whole-tree chips. The structure of the network is the same for ag waste, but it may not require drying and would start either via field harvesting or, more likely, at an intermediate processor, such as a sugar cane mill (their fuel is bagasse, which is sugarcane fiber after pressing) or nut shelling operation (e.g., pecans or walnuts).

For example, the network indicates the steps required to process the raw source (in the diagram, wood is used as the raw material) from the source site. Each block indicates a process in-

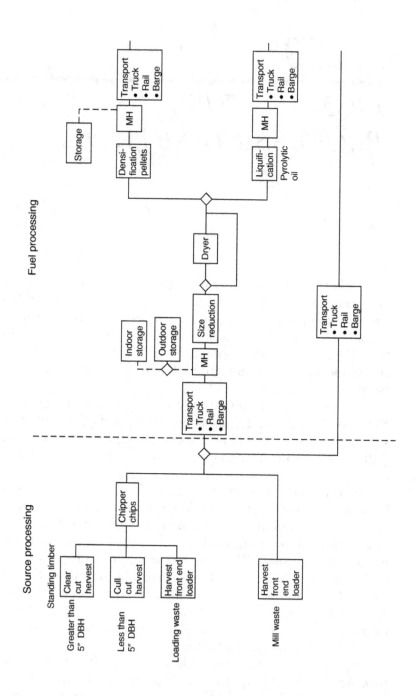

Source processing

Fuel processing

Standing timber

Greater than 5" DBH — Clear cut hervest

Less than 5" DBH — Cull cut harvest

Loading waste — Harvest front end loader

Mill waste — Harvest front end loader

Chipper chips

Transport
• Truck
• Rail
• Barge

MH — Size reduction

Indoor storage

Outdoor storage

Dryer

Densification pellets

Storage

MH — Transport
• Truck
• Rail
• Barge

Liquification — Pyrolytic oil

MH — Transport
• Truck
• Rail
• Barge

Transport
• Truck
• Rail
• Barge

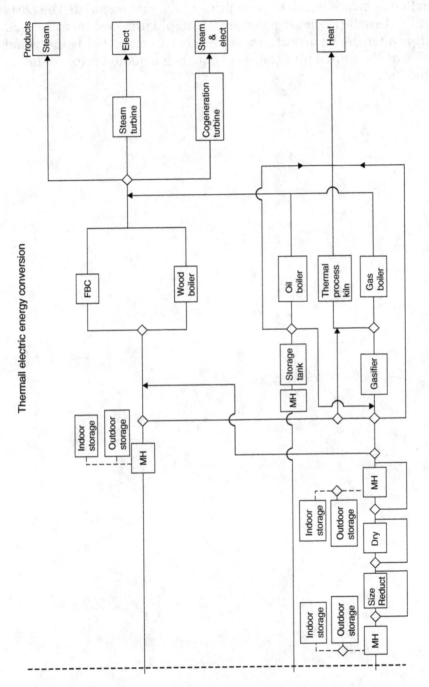

**Figure 13-1.** Wood energy processing network.

223

volving capital equipment and manpower. Each process has a cost that must be added to the base price of the raw material. The costs build at each processing step, for example, the cost of waste and chips after the transport process has been completed is the wood price *at the source* plus transport equals the quoted price for *delivered* fuel.

# CHAPTER 14

# EXAMPLE FEASIBILITY STUDY: NONFOREST PRODUCTS FACILITY

This chapter is a case study of a feasibility study performed by Georgia Tech engineers for a wood fired boiler at a textile manufacturing facility in 1980. It is presented here as a guide and template for consultants, engineers, and business planners to carry out similar evaluations on midsized nonforest product facilities. This study led to the successful installation of a wood fired boiler at the subject textile mill in Georgia.

## BACKGROUND

The textile mill manufactures yarn for the carpet industry. It employs 450 people and supplies products to domestic and international customers. The plant uses steam from the boilers to "heat set" and dry the carpet yarn in order to provide a permanent twist to the yarn. This process is accomplished in autoclaves and steam dryers, and, in winter, steam is also used for space heating. Formerly, the plant boilers ran on natural gas and oil; approximately 85% of the energy consumption was provided by natural gas and 15% by No. 2 fuel oil.

Initial visits by the engineering team found that the plant was a good prospect. The potential for success was based on the following considerations:

- A preliminary economic analysis performed on the basis of the energy usage data indicated that a wood energy system would

be economically feasible and that the textile mill had a strong potential for conservation of scarce gas and oil fuels.
- Meetings with the engineering and plant personnel revealed that the textile mill had a competent technical staff interested in the conversion to wood.
- A survey of the local wood supply situation indicated that wood fuel in the form of sawdust, bark, and wood chips would be available at economical prices.
- Financial records of the textile mill revealed that the textile mill was stable and the management had a positive attitude with respect to participating in a demonstration project.

## CONCEPTUAL DESIGN

The primary goals of the conceptual design study of the proposed wood energy system were to determine boiler size, determine the wood fuel storage volume required, determine the optimum wood handling system, and develop plans for the plant layout.

In order to perform this task, information on the energy use of the existing plant, specifications of the existing boilers, characteristics of the steam consuming equipment, and available space for the installation of the wood system was collected. The engineering staff and plant personnel of the textile mill provided a considerable amount of information and assistance in these efforts. Information on the existing boiler plant is summarized in Table 14-1. Monthly energy consumption data is shown in Table 14-2, and a plot plan of the area adjacent to the boiler house is shown in Figure 14-1.

### Size of the Boiler

In sizing the boiler, the existing energy consumption, the size and utilization of the existing boilers, characteristics of the steam consuming equipment, and the future plans of the textile mill were taken into consideration. There were no steam flowcharts to indicate the total instantaneous demand on the boilers. Steam flowcharts connected with the major steam consuming equipment such as the autoclaves and steam dryers indicated the cyclical nature of the steam demand by the equipment. Discussions were held with the engineering and plant personnel concerning the

**Table 14-1.** Feasibility study background data

| Parameter | Comment |
|---|---|
| Product Line: | "Heat set" nylon carpet yarn for the carpet industry |
| Number of employees | 450 |
| Boiler plant information | There are three boilers with descriptions as follows |
| Boiler No. I | "Eclipse" package boiler |
| Date of installation | 1972 |
| Capacity | 200 hp (6900 lb steam/hr) |
| Pressure | 100 psig |
| Fuel | Gas or oil |
| Emission control | NA |
| Ash handling | None |
| Boiler efficiency | 82.5% (approximately) |
| Boiler No. 2 | "Eclipse" package boiler |
| Date of installation | 1972 |
| Capacity | 350 hp (12,000 lb steam/hr) Other specifications are the same as those of Boiler No. 1. |
| Boiler No. 3 | "Cole" HRT boiler (used as standby) |
| Date of installation | 1920 |
| Capacity | 100 hp |
| Fuel | Gas or oil |
| Steam usage | Steam is used in autoclaves for the "heat set" process. This process provides a permanent twist to the yarn. During winter, approximately 10% of the space heating demand is met by steam. |
| Plant operation | 24 hours/day, 6 days/week, 50 weeks/year |
| Energy consumption | Natural gas: 51,934 × 103 cu ft/year |
|  | No. 2 fuel oil: 53,464 gals/year |
| Other information | There is sufficient yard area available for wood handling, storage, and installation of the wood-fired boiler. |

loading of the existing boilers and the addition of another dryer. Based on these considerations, a 400 hp (13,000 lb steam/hr) boiler was judged to be adequate by the engineering and plant personnel.

## Wood Fuel Storage and Handling

To estimate the size of the storage and the wood handling equipment, assumptions were made with regard to wood fuel properties and system efficiency (Table 14-3).

A storage silo of this size, or a covered storage shed of floor area of approximately 2600 ft$^2$ (assuming average of 6 ft of pile

**Table 14-2.** Energy consumption for textile mill

| Period | Month | Natural Gas | | Oil | | Total Btu |
| --- | --- | --- | --- | --- | --- | --- |
| | | Therms | Btu | Gals | Btu | |
| 4/28–5/30/78 | May 78 | 49,955 | $4.9955 \times 10^9$ | | | $4.9955 \times 10^9$ |
| 5/30–6/28/78 | June 78 | 45,093 | $4.5093 \times 10^9$ | | | $4.5093 \times 10^9$ |
| 6/28–7/28/78 | July 78 | 36,710 | $3.6710 \times 10^9$ | | | $3.6710 \times 10^9$ |
| 7/28–8/29/78 | Aug 78 | 46,868 | $4.6868 \times 10^9$ | | | $4.6868 \times 10^9$ |
| 8/29–9/29/78 | Sept 78 | 41,625 | $4.1625 \times 10^9$ | | | $4.1625 \times 10^9$ |
| 9/29–10/31/78 | Oct 78 | 52,253 | $5.2253 \times 10^9$ | | | $5.2253 \times 10^9$ |
| 10/31–12/1/78 | Nov 78 | 48,322 | $4.8322 \times 10^9$ | | | $4.8322 \times 10^9$ |
| 1211/78–1/3/79 | Dec 78 | 35,782 | $3.5782 \times 10^9$ | 8,000 | $1.1200 \times 10^9$ | $4.6982 \times 10^9$ |
| 1/3/79–1/31/79 | Jan 79 | 27,742 | $2.7742 \times 10^9$ | 23,215 | $3.2501 \times 10^9$ | $6.0243 \times 10^9$ |
| 1/31/79–3/1/79 | Feb 79 | 34,466 | $3.4466 \times 10^9$ | 14,749 | $2.0649 \times 10^9$ | $5.5115 \times 10^9$ |
| 3/1/79–4/1/79 | Mar 79 | 55,409 | $5.5409 \times 10^9$ | 7,500 | $1.0500 \times 10^9$ | $6.5909 \times 10^9$ |
| 4/1/79–4/30/79 | April 79 | 45,123 | $4.5123 \times 10^9$ | | | $4.5123 \times 10^9$ |
| Annual consumption | | | | | | $59.4198 \times 10^9$ |

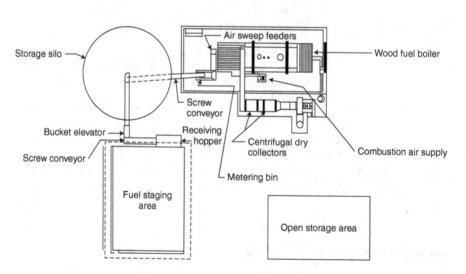

**Figure 14-1.** Wood-fired boiler plant.

**Table 14-3.** Fuel properties and system efficiency

| Parameter | Comment |
|---|---|
| Wood fuel | Sawdust, bark, and chips 50% moisture content |
| Heat value | 4,000 Btu/lb |
| Bulk density | 24 lb/ft³ |
| System thermal efficiency | 65% |
| Heat input required to generate 1 lb of steam | 1000 Btu |
| Calculations | |
| Output of boiler at rated capacity | 400 hp × 34.5 lb-steam/hr (boiler hp) × 1000 Btu/lb-steam = 13.8 × 106 Btu/hr |
| Wood fuel consumption/hr | 13.8 × 106 Btu/hr/4000 Btu/lb × 0.65 = 5,308 lb/hr |
| Volume flow rate of wood | 221 ft³/hr |
| Storage volume for covered storage, 72 hours (rated capacity) | 15,912 ft³ |

depth) can be used. A concrete silo was the final choice. Storage volume calculations are shown in Table 14-4.

## Wood Fuel Handling

For a plant of this size, preliminary feasibility studies indicated that an automatic fuel receiving hydraulic truck dump would be an expensive alternative. Simpler systems such as the use of a front-end loader or fuel delivery by live-bottom delivery vans were considered to be economical alternatives.

**Table 14-4.** Storage volume calculations

| | |
|---|---|
| Open storage area Concrete floor area for 28 days storage, assuming average of 12 ft pile depth | 12,376 ft² |
| Wood Requirements | 2.654 tons/hr |
| Wood required/hour | = 3 truckloads/day at rated capacity = 2 truckloads/day at 60% to 70% of rated capacity |
| Wood required/year at 100% capacity | 19,000 tons/year |
| Wood required/year at 70% capacity | 13,300 tons/year |

Other fuel handling equipment such as conveyors and elevators designed to handle the design volume of wood (221 ft³/hr) would be included in the system.

## Space Requirements

Based on manufacturer's data for wood-fired boilers, wood handling equipment, and our own calculations, it was found that a 400 hp wood system including a storage silo would require approximately 7200 ft² of space for the boiler house, silo fuel unloading, and driveway for trucks to enter the fuel staging area.

## WOOD FUEL AVAILABILITY

As part of the feasibility study, inquiries were made regarding wood fuel availability near the plant. Wood fuel suppliers were contacted both by Georgia Tech personnel and by the staff of the textile mill. Four wood fuel suppliers located near the plant showed great interest in developing a steady wood supply.

Additionally, the project on wood fuel sources and suppliers conducted by the University of Georgia as part of the feasibility studies for this project resulted in a list of additional suppliers in several counties close to the textile mill.

## CONTRACTOR SELECTION

Based on our discussions with the engineering staff of the textile mill and the study of the fuel supply situation, it was decided that the selected boiler should be capable of burning sawdust, bark, wood chips, and other kinds of wood waste with up to 50% moisture content on a wet basis. Various boiler types were considered. On the basis of dependability, performance, fuel capability, and first cost, an HRT (horizontal return tube) boiler was considered to be a good choice. A number of boiler manufacturers were contacted based on experience with previous projects. Five potential suppliers visited the plant site.

During the visits, the representatives from these firms investigated the details of the existing boiler plant, including space availability and use. Some of the vendors explained the details of the systems they were going to propose, including details of similar systems supplied to other companies. Based on their examinations of the plant site and the requirements of the proposed wood energy plant, four of the companies submitted proposals. There were some variations in the components of the systems proposed by these firms. The equivalent costs of the complete systems varied from approximately $1,200,000 to $1,600,000 (in 2008 dollars, 2.6 CPI inflator, 1980–2008). In the selection of the contractor, the following factors were taken into consideration:

- The cost of the wood energy system quoted by the vendor
- The type of wood boiler and the details of the wood energy system
- The reliability and the service that would be furnished by the vendor, as determined by previous experience and reputation
- Contractors' guarantees on system performance and emission characteristics

The final vendor choice was made after the textile mill vice president for engineering visited a similar installation provided by that vendor. The accepted proposal was the low bid among those submitted.

The project was done on a turnkey basis, with all of the functions of design, manufacture, installation, and startup of the entire wood system borne by the contractor. The only exceptions were the boiler house and foundation, the sheltered fuel staging area, and the open storage area.

## Contract Document Preparation

The proposal included details of the wood energy system and budget prices for the various components. The proposal was reviewed by the textile mill plant personnel and feasibility study engineers with respect to the costs of the system components, the optional equipment, and the system layout. Meetings were also held with representatives from the vendor and the final contract included the following items.

*Fuel Storage System*
- Poured concrete silo
- Supreme unloader

*Fuel Feeding System (Unloader to Boiler)*
- Conveyor—unloader to metering bin
- Metering bin
- Transfer conveyor—metering bin to air sweep feeders
- Air sweep feeders
- Fuel receiving area, sheltered

*Steam Generating System*
- Industrial horizontal tubular boiler, Model 3-2600-158
- Cast iron pinhole grates
- Combustion air system
- Steel casing
- Refractories
- Rear arch
- Prewired automatic control panel
- Primary air pollution control equipment

*Exhaust Gas Handling System*
- Exhaust gas duct
- Secondary air pollution and control equipment
- Clean gas duct
- Draft inducer (fan)
- Exhaust gas stack

*Auxiliary Equipment and Services*
- Boiler feed water system
- Chemical feed system
- Steam flow meter
- Piping material for installation of the system within boiler room
- Electrical materials for field wiring of system
- Labor for installation of system

The boiler house and foundation, the sheltered fuel staging area, and the open storage area were built by a local construction company under a separate purchase order from the textile mill.

## SYSTEM DESCRIPTION

The system (as shown in Figure 14-1) consists of the following major components:

- Horizontal return tube boiler (HRT)
- Wood storage
- Wood fuel handling

Each one of these components is described briefly in the following sections.

### Horizontal Return Tube Boiler

The HRT boiler is a firetube boiler designed to produce 13,320 lbs/hr of saturated steam (400 hp) at 130 psig from feed water supplied to the unit at 220°F. The boiler is designed to burn wood fuel with up to 50% moisture content (wet basis) such as sawdust, bark, and chips. Wood fuel enters the grate area by means of air sweep feeders.

### Wood Storage

A slip-form poured concrete silo, 24 ft diameter by 48 ft high, provided with a Supreme unloader, is used for storage. The usable storage capacity of this silo is 14,464 cu ft and is sufficient to store wood fuel for approximately 72 hours when the boiler is operating at 70% load. For long-term storage, an open storage area of approximately 10,000 ft$^2$ provided with a concrete floor and sufficient to store wood for approximately 28 days has been built on adjacent land owned by the textile mill.

### Wood Fuel Handling

Wood fuel is supplied to the plant by means of live-bottom trailers. They can unload the fuel in the open storage area or in the fuel staging area, which is located close to the wood storage silo. The wood fuel in the fuel staging area is transferred to the receiving hopper located in the corner by means of a front-end loader. The receiving hopper is approximately 4 ft wide, 8 ft long, and 6.5 ft high, and the fuel received in this hopper is delivered to the storage silo by means of a screw conveyor and bucket elevator. Fuel is transferred from the silo to the metering bin and from the metering bin to the air sweep feeders by means of screw conveyors.

## FEASIBILITY STUDIES

A preliminary economic analysis was performed on the basis of budget estimates for the equipment and wood fuel costs. This was done prior to the acceptance of the textile mill as a candidate for the demonstration project; partial funding was available from a DOE grant. A refined feasibility study based on the quoted prices for the equipment was later prepared. Addressed were major items such as economic analysis and the wood supply situation.

### Economic Analysis

The economic analysis includes the costs associated with the wood system and the existing system, the first year savings, and the payback period. Life cycle costs based on escalation rates for the costs and the present worth savings over the life of the equipment were calculated using a computer. The items used in the analysis included:

- Total costs (including operating costs, etc.)
- Total annual savings due to the installation of the wood system
- Approximate payback period
- Life cycle costs and savings, allowing for escalation rates

### Costs

Costs were based on the information obtained from the manufacturers, plant engineering personnel, and discussion with professionals in the field. Current prices of conventional fuel were obtained from the textile mill.

In computing the capital cost, an allowance for the federal investment and alternative energy property tax credits amounting to 20% of the investment is made. Additionally, the DOE grant to the textile mill toward the demonstration is subtracted to arrive at the final adjusted capital investment, with adjusted costs shown below (2008 dollars):

| | |
|---|---|
| Cost of the system: | $1,200,000 |
| Less 20% federal tax credits | $240,000 |
| Less DOE Grant | $380,000 |
| Adjusted Capital | $580,000 |

The system includes the boiler and related equipment described earlier: building, storage, and so on. In arriving at the nonfuel operating costs and fuel costs, the following data and assumptions are used.

*Operating Costs*
  a. Wood system:
       Capacity = 400 hp or 13,800 lbs steam/hr
       Wood fuel = Sawdust, bark, whole tree chips
       Moisture content = 50% (wet basis)
       Heat value = 4000 Btu/lb
       Wood system thermal efficiency = 65%
  b. Existing system:
       Fuel consumption = 87.4% natural gas
                        = 12.6% No. 2 oil
       Natural gas = 51,394,400 ft$^3$/year
       No. 2 fuel oil = 53,464 gals/year
       Heat value of natural gas = 1,000 Btu/ft$^3$
       Heat value of No. 2 oil = 140,000 Btu/gal
       System efficiency = 80% (gas–oil)
  c. Other assumptions:
       Life of wood system = 20 years
       Interest rate = 12%
       Cost of electricity = $0.078/kWh
       Labor = $14.60/hour
       Capital of existing system = Fully paid
  d. Fuel costs:
       Annual energy consumption = 59.42 × 10$^9$ Btu/year (gas
                                   and oil)
       Net energy output = 47.536 × 10$^9$ Btu/year (80% efficiency)
       Wood fuel consumption = 47.536 × 109/4000 × 2000 × 0.65
                               tons/year
                             = 9142 tons/year
         Cost of wood fuel @ $21/ton = 9142 × 21 = $190,000/year
                               (in 2008 $)
       Conventional fuel costs (2008 dollars)
         Cost of No. 2 fuel oil = 53,464 gallons × $2.24/gal
                               = $120,000/year
         Cost of natural gas = 51,934.8 MCF × $7.80/MCF
                             = $405,000/year
         Total cost of conventional fuels = $525,000/year
                               (in 2008 $)

## Life Cycle Analysis

The costs above are useful for evaluating the savings for the first year only (circa 1980–1981). In order to assess the economics of this project on a long-term basis, the effects of specific cost increases and inflation have to be taken into consideration. Therefore, life cycle costs were developed by using escalation rates for the various costs. The following annual escalation rates were used for the various costs at that time:

| | |
|---|---|
| Operating costs (wood system) | 7% |
| Operating costs (conventional system) | 7% per year |
| Property tax and insurance | 6% per year |
| Wood fuel | 9% per year |
| Gas and oil | 17% per year for the first 10 years |
| | 10% per year for the second 10 years |
| Coal | 9% per year |

Net present worth calculations showed the savings in 2008 dollars to be $180,000 for the first year, and $320,000 per year for 5 years, and $500,000 per year for 10 years. The payback period, with the DOE grant and Federal investment tax credit figured in, was 3 years.

## OPERATIONS

In the first 6 months of operation, material handling was the only minor problem experienced. This was due to oversize fuel binding up a feed auger. To prevent oversize fuel (mill residue) from entering the feed auger, plant personnel designed a shaker screen. It was constructed by plant mechanics, and it was placed on the front end of the fuel handling train. The screen was sized to allow only fuel of 2″ or less to pass through, and this addition eliminated auger jams.

Accumulated fly ash and residues are manually cleaned from the HRT boiler on weekends when the boiler was shut down.

An unforeseen benefit of the new boiler installation (most probably attributable to the replacement of two fossil-fuel boilers operating in parallel with a single wood-fired boiler) was a time reduction of approximately 10% for the two plant processes that require steam. It is likely that the wood-fired boiler had both more steam space above the water line and more retained heat in the steel and refractory, allowing it to take short-duration steam loads more easily.

The visual stack opacity reading was estimated to be 8%, well under the 20% limit established by the state.

The installation performed as dependably as the fossil-fuel system it replaced.

# EQUIPMENT MANUFACTURERS/VENDORS LISTING

**1-1.** Fuel Preparation, Handling, Storage, Transport, and Related Equipment

| Company | Products/Services |
|---|---|
| 1 Advanced Metal Fabricators—Filtrex (formerly Steelcraft Corporation) Arlington, TN Phone: 901-867-7814 amffiltrex.com | High/low pressure pneumatic material conveying systems, filter collectors, storage bins |
| 2 Aeroglide Corporation Raleigh, NC Phone: 919-851-2000 www.aeroglide.com | Dryers |
| 3 Air-O-Flex Equipment Company Roseville, MN Phone: 651-631-2527 airoflex.qwestoffice.net | Truck and rail dumps |
| 4 Andritz Sprout-Bauer Muncy, PA Phone: 570-546-8211 www.andritzsproutbauer.com | Equipment for pelletizing, size classification, size reduction, and materials handling |
| 5 Atlas Systems Corporation Spokane, WA Phone: 509-535-7775 www.atlassystems.nets | Shredded wood residue storage silos, automatic discharge system |
| 6 California Pellet Mill Co. Crawfordsville, IN Phone: 765-362-2600 www.cpmroskamp.com/pelletmill | Pelletizer equipment |

*(continued)*

**1-1.** Fuel Preparation, Handling, Storage, Transport, and Related Equipment *(Continued)*

| | Company | Products/Services |
|---|---|---|
| 7 | Earth Care Products, Inc. (formerly Guaranty Performance Co., Inc.) Independence, KS Phone: 620-331-0090 www.ecpisystems.com | Rotary dryers, related fuel-handling equipment |
| 8 | Ederer, LLC (PaR Systems) Seattle, WA Phone: 206-622-4421 www.par.com | Rake cranes, conveyors |
| 9 | Eriez Magnetics Erie, PA Phone: 814-835-6000 www.eriez.com | Metal separators |
| 10 | Fulghum Industries, Inc. Wadley, GA Phone: 478-252-5223 www.fulghum.com | Tree shears, wood chippers, screens, sawmills |
| 11 | Hallco Manufacturing Co., Inc. Tillamook, OR Phone: 800-542-5526 www.hallcomfg.com | Live-bottom trailers |
| 12 | Harvey Manufacturing Corporation Beebe, AR Phone: 501-882-6000 www.harveymfg.com | Fuel storage, handling, and preparation systems |
| 13 | Industrial Process Equipment Geneva, IL www.industrialprocessequipment.com | Bulk solids conveying, processing, storage, and weighing solutions |
| 14 | John Deere Corporation Moline, IL Phone: 309-765-8000 www.deere.com | Crop residue densifiers, feller bunchers, skidders, harvesters |
| 15 | Kinergy Corporation Louisville, KY Phone: 502-366-5685 www.kinergy.com | Vibrating screens, feeders, conveyors, dust screens |
| 16 | K-Tron International Pitman, NJ Phone: 856-589-0500 www.ktroninternational.com | Solids handling equipment, metering conveyors |

**1-1.** Fuel Preparation, Handling, Storage, Transport, and Related Equipment *(Continued)*

| Company | Products/Services |
| --- | --- |
| 17  Laidig, Inc.<br>Mishawaka, IN<br>Phone: 574-256-0204<br>www.laidig.com | Wood refuse handling systems |
| 18  Maren Engineering Corporation<br>South Holland, IL<br>Phone: 708-333-6250<br>www.marenengineering.com | Baling press for wood shavings, sawdust, hydraulic balers, shredders |
| 19  M-E-C Company<br>Neodesha, KS<br>Phone: 620-325-2673<br>www.m-e-c.com | Rotary drum dryers, flash tube dryers, wood residue fuel preparation systems |
| 20  Morbark, Inc.<br>Winn, MI<br>Phone: 989-866-2381<br>www.morbark.com | Fuel harvesting machinery, chip classification hardware, wood chipping and grinding equipment |
| 21  Munson Machinery Company, Inc.<br>Utica, NY<br>Phone: 800-944-6644<br>www.munsonmachinery.com | Hogs, hammermills, wood chipping/shredding, size reduction equipment |
| 22  Precision Husky Corporation<br>Leeds, AL<br>Phone: 205-640-5181<br>www.precisionhusky.com | Total-tree chippers, screens, conveyors |
| 23  Lacey-Harmer Company<br>(formerly Rens Manufacturing Co)<br>Portland, OR<br>Phone: 800-367-9992<br>www.laceyharmer.com | Metal detectors |
| 24  Rexnord, Inc.<br>Louisville, KY<br>Phone: 866-REXNORD<br>www.rexnord.com | Vibrating conveyors and conveyor accessories |
| 25  Schutte-Buffalo Hammer Mill<br>Buffalo, NY<br>Phone: 716-855-1555<br>www.hammermills.com | Wood grinders, air conveyors, screw elevating equipment, dumps, and hoists |
| 26  Screw Conveyor Corporation<br>Hammond, IN<br>Phone: 219-931-1450<br>www.screwconveyor.com | Wood conveyors, truck dumps |

*(continued)*

**1-1.** Fuel Preparation, Handling, Storage, Transport, and Related Equipment *(Continued)*

| Company | Products/Services |
|---|---|
| 27 Triple/S Dynamics, Inc.<br>Dallas, TX<br>Phone: 800-527-2116<br>www.sssdynamics.com | Conveyors, sizing equipment, materials handling, material screening |
| 28 Wellons, Inc.<br>Vancouver, WA<br>Phone: 360-750-3500<br>www.wellonsusa.com | Wood-fuel storage bins, conveyors |
| 29 West Salem Machinery<br>Salem, OR<br>Phone: 800-722-3530<br>www.westsalem.com | Wood grinders, shedders, and screens |

**1-2.** Combustion and Heat Recovery

| Company | Products/Services |
|---|---|
| 1 Alstom Power<br>Windsor, CT<br>Phone: 860-285-3462<br>www.alstom.com | Industrial boilers, package boilers |
| 2 C-B Nebraska Boiler<br>(formerly ABCO Industries)<br>Lincoln, NE<br>Phone: 402-434-2000<br>www.neboiler.com | Waste burner boilers (wood, sawdust, agricultural waste) |
| 3 Babcock & Wilcox<br>Power Generation Group<br>Lynchburg, VA<br>Phone: 434-522-6000<br>www.babcock.com | Wood-fired boilers |
| 4 Biomass Combustion Systems<br>Worcester, MA<br>Phone: 508-798-5970<br>www.biomasscombustion.com | Wood-fired industrial and package boilers |
| 5 Bigelow Company<br>Clarence Center, NY<br>Phone: 716-741-1300<br>www.thebigelowcompany.com | Wood-fired boilers in firetube and watertube designs |
| 6 Coen Company, Inc.<br>Woodland, CA<br>Phone: 530-668-2100<br>www.coen.com | Wood-fired suspension and cyclone burners |

**1-2.** Combustion and Heat Recovery *(Continued)*

| | Company | Products/Services |
|---|---|---|
| 7 | Combustion Service & Equipment Co. Pittsburgh, PA Phone: 412-821-8900 www.combustionservice.com | Wood-fired boilers installation and service |
| 8 | Deltak Corporation Plymouth, MN Phone: 763-557-7440 www.deltak.com | Wood-fired and waste-to-energy boilers |
| 9 | Detroit Stoker Company Monroe, MI Phone: 734-241-9500 www.detroitstoker.com | Wood-fired boilers |
| 10 | Dorr-Oliver Eimco Alpharetta, GA Phone: 770-394-6200 www.dorrolivereimco.com | Wood-fired fluidized beds |
| 11 | Earth Care Products, Inc. (formerly Guaranty Performance Co., Inc.) Independence, KS Phone: 620-331-0090 www.ecpisystems.com | Wood-fired burners, suspension and cyclone wood burners |
| 12 | Energex Ltd. Mifflintown, PA Phone: 717-436-0015 www.energex.com | Wood-fired suspension and cyclone burners, wood pelletizing |
| 13 | Energy Products of Idaho Coeur d'Alene, ID Phone: 208-765-1611 www.energyproducts.com | Wood residue combustion and energy recovery systems, fluidized beds |
| 14 | Foster Wheeler Corporation Marietta, GA Phone: 770-509-0337 www.fwc.com | Wood residue fired boilers, industrial and utility |
| 15 | Heuristic Engineering Inc. Vancouver, BC, Canada Phone: 604-263-8005 www.heuristicengineering.com | Complete wood energy combustion systems |
| 16 | Hurst Boiler & Welding Company Phone: 229-346-3545 Coolidge, GA 37138 www.hurstboiler.com | Wood-fired boiler systems |

*(continued)*

**1-2.** Combustion and Heat Recovery *(Continued)*

| Company | Products/Services |
|---------|-------------------|
| 17  Johnston Boiler Company<br>Ferrysburg, MI<br>Phone: 616-842-5050<br>www.johnstonboiler.com | Wood-fired fluidized beds |
| 18  McBurney Boiler Systems<br>Norcross, GA<br>Phone: 888-448-6610<br>www.mcburney.com | Complete biomass boiler systems |
| 19  Ray Burner Company<br>Richmond, CA<br>Phone: 510-236-4972<br>www.rayburner.com | Packaged firetube wood residue boilers and auxiliary burners |
| 20  The Teaford Co. Inc.<br>Alpharetta, GA 30009<br>Phone: 770-475-5250<br>www.teafordco.com | Wood-fired boiler systems |
| 21  Wellons, Inc.<br>Vancouver, WA<br>Phone: 360-750-3500<br>www.wellonsusa.com | Wood-residue-fired steam generating plants, drying systems, boilers |
| 22  York-Shipley Boilers, Inc.<br>Louisville, KY<br>Phone: 800-228-8861<br>www.wareinc.com/york | Complete wood-residue-fired combustion systems or conversion units for existing boilers |

**1-3.** Biomass Gasification and Pyrolysis Systems

| Company | Products/Services |
|---------|-------------------|
| 1  Biomass Gas and Electric (BG&E)<br>Norcross, GA<br>Phone: 770-662-0256<br>www.biggreenenergy.com | Biomass gasification systems |
| 2  Energy Products of Idaho<br>Coeur d'Alene, ID<br>Phone: 208-765-1611<br>www.energyproducts.com | Single fluid-bed biomass gasifiers |
| 3  Heuristic Engineering Inc.<br>Vancouver, BC, Canada<br>Phone: 604-263-8005<br>www.heuristicengineering.com | Biomass gasification systems |

**1-3.** Biomass Gasification and Pyrolysis Systems *(Continued)*

| | Company | Products/Services |
|---|---|---|
| 4 | Nexterra Energy Corp. Vancouver, BC, Canada Phone: 604-637-2501 www.nexterra.ca | Gasification systems |

**1-4.** Electric Power Generation Equipment

| | Company | Products/Services |
|---|---|---|
| 1 | ABB, Inc. Heights, PA Phone: 724-295-6044 www.abb.us | Turnkey power generation systems, waste-to-energy systems |
| 2 | Dresser-Rand Houston, Texas Phone: 713-354-6100 www.dresser-rand.com | Steam turbine–generator sets, 0.5–100 MW |
| 3 | The Elliott Company Jeannette, PA Phone: 1-800-635-2208 www.elliott-turbo.com | Steam turbines and power generation equipment |
| 4 | General Electric Co. Schenectady, NY Phone: 518-385-2211 www.gepower.com | Power generation systems |
| 5 | Siemens Power Generation Berlin, Germany Phone: +49 180 524 7000 www.powergeneration.siemens.com | Power generation systems |

**1-5.** Pollution Control Equipment

| | Company | Products/Services |
|---|---|---|
| 1 | Aget Manufacturing Company Adrian, MI Phone: 1-800-832-2438 www.agetmfg.com | Cyclone collectors, coolant/mist collectors, baghouse systems |
| 2 | American Air Filter Co., Inc. Louisville, KY Phone: 1-800-477-1214 www.aafintl.com | Pollution control equipment |

*(continued)*

**1-5.** Pollution Control Equipment *(Continued)*

| Company | Products/Services |
| --- | --- |
| 3 Donaldson Company, Inc.<br>St. Paul, MN<br>Phone: 952-887-3131<br>www.donaldson.com | Filters and dust collection systems |
| 4 Fisher-Klosterman, Inc.<br>Louisville, KY<br>Phone: 502-572-4000<br>www.fkinc.com | Wet scrubbers, air pollution control equipment |
| 5 FMC Corporation<br>Environmental Solutions Division<br>Philadelphia, PA<br>Phone: 815-228-1306<br>www.envsolutions.fmc.com | Dust collectors, scrubbers, $NO_x$ and $SO_x$ abatement equipment |
| 6 Hamon Research-Cottrell<br>Somerville, NJ<br>Phone: 908-685-4000<br>www.hamon-researchcottrell.com | Electrostatic precipitators, baghouses, scrubbers, $NO_x$ removal, flue gas desulfurization |
| 7 Industrial Boiler and Mechanical<br>Chattanooga, TN<br>Phone: 423-629-1117<br>www.industrialboiler.com | Pollution control equipment, autofeed, controls |
| 8 McCarthy Products Company<br>Seattle, WA<br>Phone: 206-522-1700<br>http://home.comcast.net/~mpco69 | Continuous single- and multichannel moisture detectors and controls |
| 9 Phelps Fan, LLC<br>Little Rock, AR<br>Phone: 501-568-5550<br>www.phelpsfan.com | Fans and blowers |

# STATE FORESTRY COMMISSION OFFICES

**Alabama**
Alabama Forestry Commission
513 Madison Avenue
P.O. Box 302550
Montgomery, AL 36130-2550
Phone 334-240-9300
www.forestry.state.al.us

**Alaska**
Alaska Division of Forestry
550 W. Seventh Ave., Suite 1450
Anchorage, Alaska 99501-3566
Phone 907-269-8474
www.dnr.state.ak.us/forestry/

**Arizona**
Arizona State Forestry Division
1110 West Washington St #100
Phoenix, AZ 85007
Phone 602-771-1400
www.azsf.az.gov/

**Arkansas**
Arkansas Forestry Commission
3821 West Roosevelt Road
Little Rock, AR 72204
Phone 501-296-1940
www.forestry.state.ar.us

**California**
Department of Forestry & Fire
 Protection
P.O. Box 944246
Sacramento, CA 94244-2460
Phone 916-653-9449
www.fire.ca.gov/

**Colorado**
State Forest Service
Colorado State University
203 Forestry Building
Fort Collins, CO 80523-5060
Phone 303-491-6303
www.colostate.edu/Depts/CSFS/

**Connecticut**
Division of Forestry
Department of Environmental
 Protection
79 Elm Street
Hartford, CT 06106-5127
Phone 860-424-3630
http//dep.state.ct.us/burnatr/forestry/

**Delaware**
Forestry Section
Department of Agriculture
2320 South Dupont Highway
Dover, DE 19901
Phone 302-698-4500
www.state.de.us/deptagri/forestry/

**Florida**
Division of Forestry
3125 Conner Blvd.
Tallahassee, FL 32399-1650
Phone 904-488-6611
www.fl-dof.com

**Georgia**
Georgia Forestry Commission
P.O. Box 819
Macon, GA 31202-0819
Phone 478-751-3523
www.gfc.state.ga.us

**Hawaii**
Division of Forestry & Wildlife
1151 Punchbowl Street, Rm. 325
Honolulu, HI 96813
Phone 808-587-0166
www.dofaw.net/

**Idaho**
Idaho Forest Products
   Commission
Phone 208-334-3292
www.idahoforests.org/

**Illinois**
Department of Natural Resources
Division of Resource Protection
   and Stewardship
One Natural Resources Way
Springfield, IL 62702-1271
Phone 217-785-8774
http//dnr.state.il.us

**Indiana**
Division of Forestry
Department of Natural Resources
1278 E SR 250
Brownstown, IN 47220
Phone 812-358-2160
www.in.gov/dnr/forestry

**Iowa**
Bureau of Forestry
Iowa Department of Natural
   Resources
Wallace Office Bldg.
502 East 9th
Des Moines, IA 50319
Phone 515-281-4924
www.iowadnr.com/forestry

**Kansas**
Kansas Forest Service
Harold G. Gallaher Bldg.
2610 Clafin Road
Manhattan, KS 66502
Phone 785-532-3300
www.kansasforests.org

**Kentucky**
Division of Forestry
627 Comanche Trail
Frankfort, KY 40601
Phone 502-564-4496
www.forestry.ky.gov

**Louisiana**
Office of Forestry
Department of Agriculture &
   Forestry
P.O. Box 1628
Baton Rouge, LA 70821-1628
Phone 225-925-4500
www.ldaf.state.la.us/divisions
   /forestry/

**Maine**
Bureau of Forestry
Department of Conservation
State House Station #22
Augusta, ME 04333
Phone 207-289-4995
www.state.me.us/doc/mfs

## Maryland
Department of Natural Resources
Forest
Service
580 Taylor Avenue
Annapolis, MD 21401
Phone 410-260-8505
www.dnr.state.md.us/forests

## Massachusetts
Bureau of Forestry
Department of Conservation and
Recreation
www.mass.gov/dcr/stewardship
/forestry/

## Michigan
Forest, Minerals, Fire
Management
Department of Natural Resources
Stevens T. Mason Bldg.
P.O. Box 30452
Lansing, MI 48909-7952
Phone 517-373-1275
www.michigandnr.com/wood

## Minnesota
Division of Forestry
Department of Natural Resources
DNR Bldg., Box 44
500 Lafayette Road
St. Paul, MN 55155-4044
Phone 612-296-6491
www.dnr.state.mn.us/forestry/

## Mississippi
Mississippi Forestry Commission
Suite 300
301 N. Lamar Street
Jackson, MS 39201
Phone 601-359-1386
www.mfc.state.ms.us

## Missouri
Forestry Section
Department of Conservation
2901 West Truman Blvd.
P.O. Box 180
Jefferson City, MO 65102
Phone 573-751-4115
www.conservation.state.mo.us
/forest/

## Montana
Division of Forestry
Department of Natural
Resources
2705 Spurgin Road
Missoula, MT 5980-3199
Phone 406-542-4300
www.dnrc.state.mt.us/forestry/

## Nebraska
Nebraska Forest Service
109 Plant Industry Building
University of Nebraska-Lincoln
Lincoln, NE 68583-0815
Phone 402-472-5822
www.nfs.unl.edu

## Nevada
Nevada Division of Forestry
2525 South Carson Street
Carson City, Nevada 89701
Phone 775-684-2500
www.forestry.nv.gov/

## New Hampshire
New Hampshire Division of
Forests and Lands
PO Box 1856
Concord, NH 03301
Phone 603-271-2214
www.dred.state.nh.us/divisions
/forestandlands

**New Jersey**
Division of Parks & Forestry
Forestry Services
P.O. Box 404
501 East State St.
Trenton, NJ 08625
Phone 609-292-2531
www.state.nj.us/dep/forestry
/service

**New Mexico**
Forestry Division
New Mexico Energy, Minerals &
Natural Resources Department
P.O. Box 1948
Santa Fe, NM 87504-1948
Phone 505-476-3325
www.emnrd.state.nm.us/forestry/

**New York**
New York State Department of
Environmental Conservation
Lands & Forests
625 Broadway
Albany, NY 12233-4250
Phone 518-402-9405
www.dec.ny.gov

**North Carolina**
Division of Forest Resources
1616 Mail Service Center
Raleigh, NC 27699-1616
Phone 919-733-2162
www.dfr.state.nc.us/

**North Dakota**
North Dakota Forest Service
P.O. Box 604
Lisbon, ND 58054
Phone 701-683-4323
www.ndsu.edu/ndsu/lbakken
/forest/NDFSHome.htm

**Ohio**
Division of Forestry
Department of Natural Resources
Fountain Square
Columbus, OH 43224
Phone 614-265-6703
www.hcs.ohio-state.edu/ODNR
/Forestry.htm

**Oklahoma**
Division of Forestry
Department of Agriculture, Food,
and Forestry
P.O. Box 528804
Oklahoma City, OK. 73152-9913
Phone 405-522-6158
www.oda.state.ok.us/aghome.htm

**Oregon**
Department of Forestry
2600 State Street
Salem, OR 97310
Phone 503-945-7200
www.odf.state.or.us/

**Pennsylvania**
Bureau of Forestry
Department of Conservation and
Natural Resources
Rachel Carson State Office Building
400 Market Street
P. O. Box 8552
Harrisburg, PA 17105-8552
Phone 717-787-6460
www.dcnr.state.pa.us/forestry/

**Rhode Island**
Division of Forest Environment
Arcadia Headquarters
260 Arcadia Road
Hope Valley, RI 02832
Phone 401-539-2356
www.dem.ri.gov/programs/bna-
tres/forest/index.htm

**South Carolina**
Forestry Commission
P.O. Box 21707
Columbia, SC 29221
Phone 803-896-8800
www.state.sc.us/forest

**South Dakota**
Division of Resource Conservation
  and Forestry
Department of Agriculture
523 E. Capitol Avenue
Pierre, SD 57501-3182
Phone 605-773-3623
www.state.sd.us/doa/forestry

**Tennessee**
Division of Forestry
Department of Agriculture
Ellington Agriculture Center
Box 40627, Mel Rose Station
Nashville, TN 37204
Phone 615-837-5431
www.state.tn.us/agriculture
  /forestry

**Texas**
Texas Forest Service
P.O. Box 310
Lufkin, TX 75902-0310
Phone 936-639-8180
http//txforestservice.tamu.edu/

**Utah**
Division of Forestry, Fire &
  State Lands
1594 West North Temple,
  Suite 3520
Box 145703
Salt Lake City, Utah 84114-5703
Phone 801-538-5555
www.ffsl.utah.gov

**Vermont**
Department of Forests, Parks, &
  Recreation
Agency of Natural Resources
103 South Main Street
Building 10 South
Waterbury, VT 05671-0601
Phone 802-241-3678
www.state.vt.us/anr/fpr/forestry

**Virginia**
Department of Forestry
900 Natural Resources Drive
Suite 800
Charlottesville, VA 22903
Phone 434-977-6555
www.dof.state.va.us

**Washington**
Department of Natural
  Resources
P.O. Box 407046
Olympia, WA 98504-7046
Phone 206-902-1650
www.dnr.wa.gov/

**West Virginia**
Division of Forestry
1900 Kanawha Boulevard East
Charleston, WV 25305
Phone 304-558-2788
www.wvforestry.com/

**Wisconsin**
Division of Forestry
Department of Natural Resources
One Gifford Pinchot Drive
Room 130
Madison, WI 53726-2398
Phone 608-231-9333
www.dnr.state.wi.us/org/land
  /forestry/

**Wyoming**
Forestry Division
Office of State Lands and
   Investments
1100 West 22nd Street
Cheyenne, WY 82002
Phone 307-777-7586
http//lands.state.wy.us/forestry.htm

# GLOSSARY

**ACFM.** Actual cubic feet per minute. The measured flow rate at process conditions as opposed to a flow rate that has been adjusted to standard temperature, pressure, and humidity.

**Ag feedstock.** Agricultural feedstock

**Air sweep feeder.** Device that uses air to transport fuel to the furnace and distribute it.

**Auger.** A screw encased in a tube or trough used for moving material.

**Baghouse.** A chamber fitted with fabric filters whose purpose is to collect solid material in the flue gas.

**Belt conveyor.** A device for transporting material, consisting of a flat, continuous belt.

**Boiler, HRT.** Horizontal return tube boiler. A boiler in which the water to be vaporized is housed in a drum through which horizontal tubes (firetubes) are run. Gases from an external furnace are passed through the tubes.

**Boiler, radiant.** A boiler in which the furnace walls are watertubes and steam is generated largely by radiant heat (as opposed to heat from convection of gases around the tubes).

**Bottom ash.** That ash that remains in, on, or beneath the grates after burning.

**Btu.** An abbreviation for "British thermal unit"; the amount of heat that is required to raise one pound of water one degree Fahrenheit.

**Bucket elevator.** A series of buckets connected by flexible links used to transport material, usually to lift material to a higher elevation.

**Bulk density.** The average density found in a large volume of material. Expressed in weight/unit volume (i.e., lb/cu ft).

*Biomass and Alternate Fuel Systems.* Edited by McGowan, Brown, Bulpitt, Walsh

**Calcining.** The heating of inorganic materials to a high temperature to drive off volatile matter or effect changes such as oxidation or change in crystalline structure.

**Carbon monoxide (CO).** A toxic gas resulting from incomplete combustion of carboniferous fuel.

**Carbon, activated.** A highly adsorbent powdered or granulated carbon.

**Carbon, fixed.** Carbon not driven off by heating to 1700°F.

**Chain or dragchain conveyor.** A series of flights or buckets connected by links so as to form an endless belt or ladder-like chain whose purpose is to move material from one point to another.

**Char.** A combustible residue from organic material—charcoal.

**Cogeneration.** The simultaneous production of electricity and thermal energy.

**Combustion efficiency.** The theoretical heat energy available from a given fuel compared to the actual heat energy made available by burning that fuel.

**Condenser.** A heat exchange device in which gas is changed to a liquid through the removal of heat.

**Condensing steam turbine.** A turbine in which steam exhaust is condensed.

**Conveyor.** A mechanical apparatus for carrying bulk material from place to place, such as an endless moving belt or a chain conveyor.

**Cubes.** Densified, cube-shaped wood, similar in size to some varieties of animal feeds.

**Cyclone.** A device for removing dust from gas by centrifugal action.

**Cyclone burner.** A device in which dry, fine particles of fuel are mixed with combustion air and blown tangentially through a series of manifolds into a cylindrical firebox.

**Densification.** A process whereby wood is mechanically compressed and the density of the material increased.

**Dry scrubber.** A device that traps the particulate matter in a gas in a moving bed of granular material. The trapped particulates may then be removed from the media by recycling.

**Dryer, cascade.** A device in which material is dried by falling through streams of hot gas.

**Dryer, flash type.** A device consisting of several loops of ducting, in which wet material and hot flue gases mingle and drying occurs.

**Dryer, flue gas.** A dryer that utilizes the exhaust gases of a furnace or boiler.

**Dryer, rotary.** A rotating cylinder through which solids to be dried and hot gases are passed.

**Dryer, tunnel.** A unit in which solids are progressively dried by being moved through a tunnel in contact with hot gases.

**Electrostatic precipitator.** A device that ionizes the particles in the gas as they enter the unit and then traps the charged particles by attracting them to oppositely charged collection plates. The plates are cleaned periodically by mechanical rapping and the dislodged particles are then collected in a hopper. In wet precipitators, particles are collected by flowing water.

**Emissions (with reference to fuel combustion).** The constituents that make up the total of the exhaust products.

**Ethanol.** Grain alcohol or "drinking" alcohol.

**Fines.** Very small particles of material, such as very fine sand or very small pieces of bark.

**Fluidized-bed combustion.** Combustion in a furnace whose combustion chamber floor has many fine holes or a pipe grid through which underfire air is forced. This air blows through a bed of noncombustible materials such as sand or small particles of limestone. The air pressure and volume are such that the bed is kept in suspension. The bed is then heated to a temperature that will ignite the fuel to be burned and the burning of the fuel injected into the bed then maintains the bed temperature.

**Fly ash.** Fine solid particles of noncombustible ash carried out of a bed of solid fuel by the draft.

**Front-end loader.** A tractor with a bucket mounted on the front of the vehicle. The bucket is hydraulically actuated so that it may scoop up material, transport it, and then unload it.

**Furnace, rotary.** A furnace in which the hearth is shaped like a turntable and may be rotated a varying speeds.

**Furnace, shaft.** A vertical, refractory-lined cylinder in which a column of solids is maintained, and through which an ascending stream of hot gases is forced.

**Gasification.** The process of converting a carbonaceous material into (primarily) a gas.

**Grate.** A device to support the fuel while it bums.

**Grate, dump.** A grate used with a spreader stoker that can itself dump ashes into the ashpit.

**Grate, inclined.** A device that supports the fuel in other than a horizontal position (a sloping position).

**Grate, traveling.** A continuous-cleaning grate that fits over bars that in turn are attached to chains that form an endless belt.

**Hammermill.** A device used to reduce the size of material by hammering action.

**HC.** Hydrocarbons.

**Heating value, higher.** A measure of heat energy of a fuel at any specified moisture content.

**Heating value, lower.** The higher heating value less the latent heat in the moisture in the fuel and combustion products.

**Hogged material.** Material (wood) that has been processed to a specific size; wood chips.

**Live bottom (with reference to material handling equipment).** A material storage bin whose floor incorporates a device for removing or unloading the material contained in the bin.

**Magnetic separator.** A device for removing iron and steel from other material through the use of magnetic attraction.

**Mass.** The amount of matter contained in a particle.

**Material screen.** A screen used to group materials according to their size. The openings in the screen have a specific size and material that is larger than the opening will not pass through the screen.

**Metering bin.** A device that both stores material and dispenses the material at a given rate.

**Methanol.** A toxic liquid commonly known as wood alcohol.

**Multiclone.** Many small-diameter cyclone tubes arranged so that the gas is cleaned as it passes through the unit.

**MWe.** Megawatt, electrical.

**Noncondensing steam turbine.** A turbine that produces power by acting as a pressure reducer; the low-pressure turbine exhaust becomes the process steam.

**$NO_x$.** Oxides of nitrogen.

**Oil, catalytic.** Oil produced by subjecting wood to high temperature and pressure in the presence of alkaline metal or other catalysts.

**Oil, pyrolytic.** Oil produced by low-temperature, partial combustion of wood. The oil contains varying amounts of water, can be burned in modified heavy oil burners, and can be used as a chemical feedstock. Unlike fuel oils, pyrolytic oil is corrosive and contains acetic, formic, propionic, and other acids and must be stored and transported in corrosion resistant materials.

**Opacity regulations.** Regulations of the Environmental Protection Agency that are used to judge the emissions from a source by visual observation.

**Overfire air.** Combustion air that is introduced above the fuel bed.

**Particle resistivity.** The measure of a particle's ability to accept and hold an electrical charge.

**Particle strength.** The physical strength of a particle, a property that must be considered when devising a fly ash collection system.

**Particulates.** Minute separate particles.

**pH.** The concentration of hydronium ions in solution. A method of expressing acidity or alkalinity on a scale whose values are from 0 to 14, with 7 representing neutrality, numbers less than 7 indicating increasing acidity, and numbers greater than 7 indicating increasing alkalinity.

**Pile burning.** Combustion of the fuel while in mounds or piles.

**Pneumatic conveyor.** A device to transport material by using high-velocity air.

**Predrying.** Removing the moisture from wood prior to using it as a fuel, as opposed to letting the moisture be removed by vaporization as the wood is burning.

**Precipitate.** A substance separated from a solution or suspension by a chemical or physical change, usually as an insoluble solid.

**Proximate analysis.** A statement of the volatiles, fixed carbon, and ash present in a fuel as a percentage of dry fuel weight.

**Pyrolysis.** A process of burning at less than stoichiometric conditions, involving the physical and chemical decomposition of solid organic matter by the action of heat in the absence of oxygen. Products of pyrolysis may include liquids, gases, and a carbon char residue.

**Radiant heat.** Heat energy traveling as wave motion (such as light travels).

**Refractory lining.** A ceramic lining capable of resisting (and maintaining) high temperatures.

**Residence time.** The length of time the fuel or gas remains in a combustion zone.

**Scrubber.** An apparatus for removing impurities and contaminants from gases by use of water or a dry granular medium.

**Shave-off scrubber system.** This system utilizes a multiclone outlet incorporating two outlet tubes instead of one. The inner tube allows the core of cleanest air to pass through while the outer tube shaves off the perimeter of the outlet gases and recycles them in the cyclone.

**Silicate.** Metallic salt containing silicone and oxygen.

**$SO_x$.** Any of the various oxides of sulfur.

**Specific weight.** The weight of a substance as compared to some standard such as water.

**Spreader–stoker.** A stoker that throws or blows fuel into the firebox of a boiler so that it is more or less uniformly spread over the firebox grate.

**Stoichiometric condition.** That condition at which the proportion of the air to fuel is such that all combustible products will be completely burned with no oxygen remaining in the combustion products.

**Stoker, underfed.** A device that feeds the fuel to the fuel bed from below the point of air admission.

**Suspension burner.** A device to combust wood fuel that has been turbulently mixed with forced air in a stream over the main fuel bed.

**Switchgear.** Equipment for transferring electrical loads.

**Thermocouple.** Electrical device that measures temperature.

**Turbine, back pressure.** A turbine whose exhaust steam is used as process heat and the turbine work is considered, more or less, a by-product.

**Turbine, extraction.** A turbine from which partly expanded steam may be extracted for use as process heat.

**Turndown ratio.** The lowest load at which a boiler will operate efficiently as compared to the boiler's maximum design load.

**Turnkey System.** A system that is built, engineered, and installed to the point of readiness for operation by the contractor.

**Ultimate analysis.** A description of a fuel's elemental composition as a percentage of the dry fuel weight.

**Underfire air.** Combustion air that is introduced below the fuel and rises through it.

**Vibrating conveyor.** A device that moves material by vibration.

**Volatiles.** Substances that are readily vaporized.

**Wet scrubber.** An apparatus that develops an interface between a scrubbing liquid and the gas to be cleaned. Particles in the gas are trapped by liquid droplets and the liquid is then collected and removed.

**Wood chipper.** A mechanical apparatus for making wood chips.

**Wood pellets.** Processed wood that has been densified and shaped into the form of small cylinders very similar to animal feed.

# INDEX

ACFM, 253
Acid gas dew point, 8
Africa, 21
Ag feedstock, 253
Agricultural feedstocks, 19
Air sweep feeder, 253
Alcohol from biomass, 42
Alternate fuels, 1
  agricultural feedstocks, 2
  application to equipment, 7, 8
  biogas, 2
  biomass, 2
  burning of, 7
  ethanol, 2
  petroleum coke, 2
  plant and refinery gas, 2
  reclaimed oil, 2
  solid wastes, 2
  used cooking oil, 2
  used tires, 2
  wood, 2
Amazon region, 21
Ambient air standards, 159
Ash content, 20
Ash slagging, 90
Ash waste, 163
Augers, 111, 253

B100 biodiesel, 38
  burner manufacturers, 41
  burners, 41
  combustion of, 40
  manufacture of, 40
B20 biodiesel, 38
Babcock and Wilcox, 52
Baghouse, 147, 253
Bark, 106
Belt conveyor, 110, 253
Biodiesel, 20, 37
Bioenergy feedstocks, 22, 23
Bioethanol, 20
Biofuels, 22, 23
Biogas, 5
Biomass fuels, 1, 6, 7, 19, 26, 95, 193
  combustion equipment, 45
  economic analysis, 191
  heating value, 38
  liquid, 37, 38
  processing network, 221
  processing routes and economics,
    191, 193
  properties, 38
  storage and handling, 95
Biomass gasification, 193
Biomass materials, 20
Biomass resource potential, 25
Boiler types, 45
  firetube, 46
  field-erected, 47
  horizontal return tube (HRT)
  package, 47
  radiant, 253
  watertube, 46
Bottom ash, 253

*Biomass and Alternate Fuel Systems.* Edited by McGowan, Brown, Bulpitt, Walsh
Copyright © 2009 American Institute of Chemical Engineers, Inc.

Btu, 253
Bubbling fluidized beds, 79
Bucket elevator, 113, 253
Bulk density, 253
Burners, 47, 76
  cyclone, 52, 76
    Earth Care Products, Inc (ECPI),
      76
  dry wood waste, 77
    Coen Company, 77
  Dutch oven, 48
  gravity feed, 49
  heaped pile, 48
  particulate problems, 49
  suspension, 52, 76

Calcining, 254
Capital costs, 216
Carbon monoxide (CO), 254
Carbon, activated, 254
Carbon, fixed, 254
Cascade dryers, 117
Catalytic oil, 193, 256
Chain conveyor, 254
Char, 254
Clean Air Act, 137
Coal, 6
Coal boiler system, 214
Cogeneration, 125, 128, 254
  economic considerations, 128
  system size, 134
Coke, 6
Combined systems, hot gas and
    thermal oil, 71
Combustion, 7, 13
  direct, 209
  efficiency, 254
  issues, 7
  systems, 2, 88
    Heuristic Engineering, Inc., 88
  temperature, 14
  time, 14
  turbulence, 14
Combustor, 79
Condenser, 254
Condensing steam turbine, 254
Construction and demolition (C&D)
    debris, 24
Controlled melting, 91

Conveyor, 254
Cooking oil, 37
Corrosion in MSW boilers, 92
Covered storage, 107
Cubes, 254
Cyclones, 145, 254
Cyclone burner, 254
Cyclone furnace, 76

Deforestation, 21
Densification, 254
Densified wood, 194
Designs for wood fuel facilities, 120
Drag chains, 110
Dragchain conveyor, 254
Dry scrubber, 254
Dryers, 115
  cascade, 117, 254
  equipment, 116
  feasibility for boiler steam plants,
    120
  flash, 117, 254
  new Installations, 120
  retrofitting, 119
  rotary, 116, 255
  single-pass, 116
  triple-pass, 116
  tunnel, 255
Dump trucks, 100
Dutch oven, 48

Economic analysis of biomass
    combustion systems, 183
  depreciation, 184
  Investment tax credits, 185
  sensitivity analysis, 189
  tax considerations, 184
Electricity generation, 203
Electrostatic precipitators (ESPs), 151,
    255
Emissions, 138, 255
  air pollution standards, 138
  areas affecting, 140
    boiler design, 141
    boiler operation, 141
    characteristics, 143
    factors, 143
    fluidized-bed hot gas heaters, 143
    fuel, 140

suspension burning, 143
control of, 137, 144
  baghouse, 147
  costs, 152
  cyclones, 145
  electrostatic precipitators, 151
  mechanical collectors, 145
  wet scrubbers, 148
  expected, 140
  regulations, 10, 138
  types of, 138
Endangered species, 163
Environment and safety, 157
Environmental impact, 157
EPA 40 CFR 279.11, 6
Equipment manufacturers and
      vendors, 239
  biomass gasification and pyrolysis
      systems, 244
  combustion and heat recovery, 242
  electric power generation
      equipment, 245
  fuel preparation, handling, storage,
      transport, and related
      equipment, 239
  pollution control equipment, 245
Ethanol, 37, 38, 193, 255
  from grain, 42
  from wood or cellulosic feedstocks,
      43

Feasibility study, nonforest products
      facility, 225
  background, 225
  conceptual design, 226
  contract document preparation, 231
  contractor selection, 230
  economic analysis, 234
  horizontal return tube boiler, 233
  life cycle analysis, 236
  operations, 236
  size of the boiler, 226
  space requirements, 230
  system description, 232
  wood fuel availability, 230
  wood fuel handling, 229, 233
  wood fuel storage and handling, 227
  wood storage, 233
Fines, 255

Flash dryers, 117
Floodplains, 163
Fluidized-bed combustion, 255
fluidized-bed combustors, 71
  Energy Products of Idaho (EPI), 75
  Johnston Fluid Fire, 74
  York-Shipley, 74
Fluidizing blower, 79
Fly ash, 255
Fossil fuels, 2
Front-end loaders, 112, 155
Fuel costs, 1, 216
Fuel properties, 95
Fuel supply, 171
  establishing a database, 172
  mill residue, 172, 173
  mill waste, 173
  NAICS (North American Industry
      Classification System), 172
  planning, 173
  procuring biomass, 171
  State Department of Industry and
      Trade (or Economic
      Development), 172
  State Forestry Commission, 172
  USDA Forestry office, 172
  whole-tree chips, 172, 173
Fuel supply safety, 166
Fuel-switching feasibility study, 175
  economic analysis, 182
  energy requirements, 176
  existing system and site data, 180
  facility layout, 181
  operational requirements, 182
  system and fuel requirements, 179
Furnaces, 76, 79
  cyclone, 76
  rotary, 255
  shaft, 255

Gas boiler system, 215
Gasification, 255
Gasifier, 85, 193
Georgia, 24
Georgia Tech, 23
Government incentives, 125
Grates, 47, 255
  dump, 51, 255
  gravity-fed, 50

Grates *(continued)*
  inclined, 255
  moving, 50, 52
  reciprocating, 79
  rotary, 51
  sloping, 79
  traveling, 51, 255
Green whole-tree chips, 106
Gross heating value (GHV), 17
Groundwater, 164

Hammermill, 256
Heat- and mass-balance programs, 14
  free, 15
    Hauck E-Solutions, 15
    HSC Software, 15
Heat recovery, 2
Heat recuperation, 8
Heating value of fuel, 16
Heating values and costs for fuels, 3
Higher heating value (HHV), 16, 256
Hogged material, 256
Hot oil systems, 59

In-plant fuel handling, 108

Kyoto Protocol, 7

Lignin, 194
Liquid fuels, 37
Live-bottom van, 98, 256
Lower heating value (LHV), 17, 256

Magnetic separator, 256
Major pollution emitting facilities, 159
Material screen, 256
Maximum achievable control
    technology (MACT), 162
Metering bin, 256
Methanol, 44, 193, 256
Mill residues, 105
Moisture content, 28
Multiclone, 256
Multifuel firing, 2

NAICS (North American Industry
    Classification System), 172
National Ambient Air Quality
    Standards, 158

National Biodiesel Board, 40
Net heating value (NHV), 17
New Source Performance Standards
    (NSPS), 158
Nitrogen oxides, 161
Noncondensing steam turbine, 256
North America, 22
$NO_x$, 256

Oak Ridge National Laboratory, 27
Oak Ridge study, 26
Occupational Safety and Health Act
    (OSHA) (Public Law 91-596),
    164
Oil boiler system, 215
Opacity regulations, 256
Open storage, 102
OSHA, 164
Overfire air, 256

Particle resistivity, 257
Particle size, 32
Particle strength, 257
Particulates, 138, 161, 257
Petrodiesel, 37
pH, 257
Physical and chemical considerations,
    34
Pile burners, 79, 257
Pile drying, 115
Pneumatic conveyors, 113, 257
Pneumatic stokers, 51
Power generation, 125
Predrying, 257
Preparation of wood fuel, 113
Prevention of Significant
    Deterioration–Clean Air Act,
    158
Production of hot gas from biomass,
    69
  components of, 70
Properties affecting heat transfer, 63
Proximate analysis, 32, 257
Purchasing and contracting for wood
    fuels, 173
Pyrolysis, 81, 257
  oil, 193, 256
  systems, 81
    Energy Resources Co. Inc., 83

Tech-Air Corporation, 82

Railroad delivery, 100
Receiving methods, 98
Reciprocating grate, 79
Refractory lining, 257
Residence time, 257
Rotary dryers, 116
Runoff, 164

Safety, 164
   fire prevention program, 167
   OSHA, 164
   personal protective equipment, 166
   personnel, 164
   storage and conveying systems, 166
   storage and handling of wood chips
      and logs, 167
Sawdust, 105
Scrubber, 257
Shave-off scrubber system, 257
Silos, 107
Size reduction, 113
Slag control, 90
   empirical tests for, 91
Sloping grate furnace, 50
Slurry treatment, 92
$SO_2$, 8
Solid waste, 163
Solid Waste Management Act, 158
Solids handling, 95
South America, 21
$SO_x$, 257
Spec oil, 5
Spreader stokers, 52, 258
Steam costs, 216
Steam generation, 209
Steam turbines, 126
Stokers, 79
   underfed, 258
Sulfur oxides, 161
Surface water, 164
Suspension burner, 258
Sustainability of wood fuels, 21

Testing for fuel properties, 18
Thermal heat transfer fluids, 61
Thermal oil, 59
   reasons for ssing, 59

versus steam, 59
Thermal oil circulation system, 65
   primary circulation loop, 66
   secondary loops, 68
Thermal oil heaters, 63
   "convective" tube bundle, 64
   combination, 65
   helical coil, 63
   multiple coil, 63
Thermal oil properties, 62
Tire-derived fuel, 5
Truck dumps, 99
Tube fouling, 92
Tunnel kilns, 53
Turbine, back pressure, 258
Turbine, extraction, 258
Turndown ratio, 49, 258

U.S. Forest Service, 23, 27
Ultimate analysis, 33
Underfire air, 258
United States, 23, 26
University of Georgia (UGA), 41
Used cooking oil, 39
Used oil, 5

Vegetable oil, 39
Vibrating conveyors, 113, 258

Waste wood boiler systems, 210
Wet scrubbers, 148, 258
Wood, 95
   bark, 96
   boards, 96
   cants, 96
   dry, 95
   fuel properties, 97
   green, 95
   hogged fuel, 96
   mill yard waste, 96
   moisture content, 95
   pellets, 96
   purchase, 96
   quantity requirements, 97, 98
   receiving methods, 98
      dump trucks, 100
      live-bottom van, 98
      railroad, 100
      truck dumps, 99

Wood *(continued)*
  sander dust, 96
  sawdust, 96
  shavings, 96
  size of, 95
  storage, 100
    covered, 107
    for green whole-tree chips and/or
      bark, 106
    for sawdust and mill residues,
      105
    open, 102
    silos, 107
  whole tree chips, 96
  yard requirements, 97
Wood chip boiler system, 212
Wood chipper, 258
Wood fuel properties, 28
Wood fueled fluidized-bed combustor
    system to produce hot gas, 213
Wood gasifiers, 85
  Biomass Gas & Electric (BG&E), 87
  Energy Products of Idaho (EPI), 87
Wood pellet boiler system, 213
Wood pellets, 52, 96, 195, 258
  global pellet imports, 203

industry, 201
  market in Europe, 202
  Pellet Fuels Institute, 202
  prices, 206
  production process, 197
  raw material source, 205
  sawmill waste, 205
  Southeastern U.S. plants, 205
  Sweden, 207
  Unit energy cost, 208
  United States, 204
  Wood Pellet Association of Canada,
    204
Wood waste gasifier with gas/oil
    boiler, 215
Wood-fired package boilers, 54
  manufacturers, 54
    Biomass Combustion Systems,
      Inc., 57
    Deltak Corporation, 58
    Hurst Boiler and Welding
      Company, 56
    Industrial Boiler
      Company/Cleaver Brooks, 55
    Teaford Company, Inc., 57
    Wellons, Inc., 58